D1446264

FINANCING ENERGY PROJECTS IN EMERGING ECONOMIES

FINANCING ENERGY PROJECTS IN EMERGING ECONOMIES

Hossein Razavi, Ph.D.

PennWell Books

PENNWELL PUBLISHING
TULSA, OKLAHOMA

PennWell Publishing Company
1421 South Sheridan / P.O. Box 1260
Tulsa, Oklahoma 74101

Printed in the United States of America

1 2 3 4 5 00 99 98 97 96

For Nahid

CONTENTS

List of Executive Overviews

The Executive Overviews

- Provide a quick overview for the first-time reader.
- Serve as reference notes for the reader who has studied the text.
- Figures and tables complement the overviews and can be read independently of the text.

EXECUTIVE OVERVIEWS IN TEXT

EXECUTIVE OVERVIEW FIGURES IN TEXT

EXECUTIVE OVERVIEW TABLES IN TEXT

Objective

This book provides first-hand information and analysis regarding how multilateral, bilateral, and commercial financiers make decisions about oil, gas, and power projects. Financing energy projects in developing countries is challenging in relation to each country's business environment, government behavior, and political risk. Hence, putting together an attractive financing package requires an intimate understanding of (1) the availability of many sources of soft loans, credits, grants, tied loans, and untied loans; (2) the objectives, tendencies, and requirements of various financiers; (3) the possibilities of combining borrowing and guarantee instruments; and (4) the effective ways of approaching financiers.

This book describes the intricacies of public and private financing of energy projects and provides guidance in preparing upstream and downstream oil and gas projects as well as power generation, transmission, and distribution projects. It is divided into three parts. Part one includes two chapters and provides a brief background of project and corporate financing and the challenges of funding projects located in developing countries. Part two consists of five chapters and introduces the major sources of funding for oil, gas, and power projects. The sources discussed include global multilateral institutions; regional development banks; bilateral aid, credit, and insurance agencies; and commercial capital markets. Part three consists of four chapters and provides guidelines for preparing successful project packages. These guidelines include methods for analyzing business environments; economic and financial viability; financial structures; and environmental concerns for oil, gas, and power projects.

ACKNOWLEDGMENTS

This book is based on what I have learned working with many people from commercial banks; bilateral financing agencies; multilateral organizations; consulting and engineering firms; and public and private oil, gas, and power companies.

I am deeply indebted to these people, who are too numerous to include here. However, I would like to thank Clive Armstrong, Hossein Askari, Eric Daffern, Bertrand de Frondeville, Robin Glantz, Luis Gutiérrez, Bjørn Håmsø, Georges Khoury-Haddad, Shigeru Kubota, Dennis O'Brien, John O'Reilly, William Porter, Bent Svensson, Barbara Treat, John Treat, Elizabeth Wang and Adam Wilson for their comments on various drafts of this manuscript. Responsibility for any errors of fact or judgment is solely mine.

Finally, I want to thank Mosemarie Boyd for assistance in preparing the manuscript, Bob Tippee and Paul Wolman for their editorial work, and Patricia Hord for her graphic design.

Hossein Razavi
The World Bank
Washington, D.C.
January 1996

Abbreviations and Acronyms

ADB	Asian Development Bank
ADR	American Depository Receipt
AfDB	African Development Bank
AfDF	African Development Fund
AIC	Agency for International Cooperation (Spain)
AKA	Ausfuhrkredit-Gesellschaft (Germany)
AusAID	Australian Agency for International Development
b/d	Barrels per day
BADC	Belgium Administration for Development Cooperation
BADEA	Arab Bank for Economic Development in Africa
BAWI	Federal Office of Foreign Economic Affairs (Switzerland)
BFCE	Banque Française du Commerce Extérieur (France)
BITS	Swedish Agency for International Technology and Economic Cooperation
bl	Barrel (for petroleum, a unit of volume equal to 42 U.S. gallons)
BMZ	Bundesministerium für Wirtschaftliche Zusammenarbeit (Federal Ministry for Economic Cooperation and Development; Germany)
BOD_5	Biochemical oxygen demand (5-day test)
BOO	Build-own-operate
BOOT	Build-own-operate-transfer
BOT	Build-operate-transfer
BP	British Petroleum
BS	Bridas SAPIC (Argentina)
BTU	British thermal unit
CARIFA	Caribbean Basin Projects Financing Authority

CDC	Commonwealth Development Corporation (United Kingdom)
CESCE	Export Credit Insurance Company (Spain)
CFCs	Chlorofluorocarbons
CFD	Caisse Française de Developpement (France)
CIDA	Canadian International Development Agency
CNPC	China National Petroleum Corporation
CO	Carbon monoxide
CO_2	Carbon dioxide
COD	Chemical oxygen demand
COFACE	Compagnie Française d'Assurance pour le Commerce Extérieur (France)
Copromex	An entity under the minister of foreign trade (Belgium)
Credit-export	A revolving fund financed by private banks and public agencies (Belgium)
DANIDA	Danish International Development Assistance
dB	Decibel (on the A-scale), a unit of sound level (relative pressure)
DEG	Deutsche Investitions-und für Entwicklunggesellschaft (Investment and Development Company; Germany)
DEH	Directorate for Development Cooperation and Humanitarian Aid (Switzerland)
DGCS	Directorate General for Development Cooperation (Italy)
DGDC	Directorate General for Development Cooperation (Austria)
DGIS	Directorate General for International Cooperation (Netherlands)
DIDC	Department of International Development Cooperation (Finland)
DREE	Direction des Relations Economiques Extérieures (France)
DSCR	Debt service coverage ratio
DTI	Department of Trade and Industry (United Kingdom)
EA	Environmental Assessment
EBRD	European Bank for Reconstruction and Development
EC	European Community
ECGD	Export Credit Guarantee Department (United Kingdom)
ECU	European Community Unit
EDC	Export Development Corporation (Canada)
EDCF	Economic Development Cooperation Fund (Korea)
EEC	European Economic Community

EFC	Danish Export Finance Corporation
EFIC	Export Finance and Insurance Corporation (Australia)
EIB	European Investment Bank
EID	Export Insurance Division (Japan)
EIU	Economist Intelligence Unit (London)
EKN	Swedish Export Credits Guarantee Board
EKR	Eksportkreditraadet (Denmark)
EMF	Electromagnetic field
EMPL	Maghreb-Europe Pipeline Limited
EPC	Engineering, procurement, and construction contracts
ESMAP	Energy Sector Management Assistance Programme
ESP	Electrostatic precipitator
EU	European Union
Export-finans	Norway's export credit agency, owned by commercial banks and GEIK
FAC	Fonds d' Aide et de Cooperation (France)
FCE	Fund for Economic Cooperation (Portugal)
FEC	Finnish Export Credit Limited
FGB	Finnish Guarantee Board
FGD	Flue-gas desulfurization
FINFUND	Finnish Fund for Industrial Development
FINNIDA	Finnish International Development Agency
FSU	Former Soviet Union
GDP	Gross domestic product
GEF	Global Environment Facility
GEIK	Guaranti-Instituttet for Eksportkreditt (Norway)
GITIC	Guangdong International Trust and Investment Corporation
GNP	Gross national product
GPW	Gross product worth
GTZ	Gesellschaft für Technische Zusammenarbeit (German Technical Assistance Corporation)
H_2S	Hydrogen sulfide
IBRD	International Bank for Reconstruction and Development
ICB	International competitive bidding
ICEX	Secretary of Commerce through the Institute for External Trade (Spain)
IDA	International Development Association
IDB	Inter-American Development Bank
IEC	Institute for Economic Cooperation (Portugal)

IEF	International Energy Finance
IFC	International Finance Corporation
IIC	Inter-American Investment Corporation
IMF	International Monetary Fund
IOC	International oil company
IPP	Independent power producer
IRR	Internal rate of return
IsDB	Islamic Development Bank
JCI	Japan Consulting Institute
JExim	Japan Export-Import Bank
JICA	Japan International Cooperation Agency
JNOC	Japan National Oil Corporation
KExim	Export-Import Bank of Korea
KfW	Kreditanstalt für Wiederaufbau (Germany)
km	Kilometer
KOICA	Korean International Cooperation Agency
kV	Kilovolt
kWh	kilowatt hour
LIBOR	London Interbank Offer Rate
LNG	Liquefied natural gas
LPG	Liquefied petroleum gas
LSTK	Lump-sum turnkey contract
MCF	Thousand cubic feet
MEDIO-CREDITO CENTRALE	Instituto Centrale per il Credito a Medio Termine (Italy)
mg	Milligram (10^{-3}g)
MIF	Multilateral Investment Fund
MIGA	Multilateral Investment Guarantee Association (member of the World Bank Group)
MITI	Ministry of International Trade and Industry (Japan)
MMBTU	Million British thermal units
MOEF	Ministry of Economy and Finance (Spain)
MOF	Ministry of Finance (Japan)
MOFA	Ministry of Foreign Affairs (Japan)
MSDS	Material Safety Data Sheet
MW	Megawatt (10^6 watts)
MWe	Megawatt electric
NCM	Nederlandsche Credietverzekering Maatschapij
NDF	Nordic Development Fund

NEPA	National Environmental Protection Act (United States)
ng	Nanogram (10^{-9}g)
ng/J	Nanogram per Joule
NGL	Natural gas liquids
NGO	Nongovernmental organization
NIB	Nordic Investment Bank
Nm^3	Normal cubic meter
NNPC	Nigerian National Petroleum Corporation
NOC	National oil company
NORAD	Norwegian Agency for Development Cooperation (Norway)
NO_x	Nitrogen oxide
NPC	National Power Corporation (the Philippines)
NPV	Net present value
NPW	Net product worth
NTF	Nigerian Trust Fund
O&M	Operations and maintenance
OAU	Organization of African Unity
ODA	Overseas Development Administration (United Kingdom)
OECD	Organisation for Economic Co-operation and Development
OECF	Overseas Economic Cooperation Fund (Japan)
OEKB	Oesterreichische Kontrollbank (Austria)
OND	Office National du Ducroire (Belgium)
OPEC	Organization of Petroleum Exporting Countries
OPIC	Overseas Private Investment Corporation (United States)
PALOP	Reference to Lusophone African countries (with historic links to Portugal)
PCB	Polychlorinated biphenyls
PEA	Pronincial Electricity Authority (Thailand)
PEFCO	Private Export Funding Corporation (United States)
PIL	Project Investment Loans
PM	Particulate matter (PM_{10} = particulate matter of 10 microns or less in size)
ppm	Parts per million
PSDU	Private Sector Development Unit
PSED	Private Sector Energy Development Program (U.S. Department of Energy)
PSEDF	Private Sector Energy Development Fund (Pakistan)
PSEF	Jamaica Private Sector Energy Fund

PTT	Petroleum Authority of Thailand
PV	Present value
ROW	Right-of-way
S&P's	Standard & Poor's Rating Group
SACE	Sezione Speciale per Assicurazione del Credito all'Esportazione (Italy)
SAR	Staff Appraisal Report (The World Bank)
SAREC	Swedish Agency for Research Cooperation with Development Countries
SEC	Securities and Exchange Commission (United States)
SESCO	Sarawak Electricity Supply Corporation (Malaysia)
SIDA	Swedish International Development Authority
SO_2	Sulfur dioxide
SO_x	Sulfur oxides (a mix of SO_2 and other oxides of sulfur—for example, SO_5)
SPA	Sichuan Petroleum Administration
SPRC	Star Petroleum Refining Company (Thailand)
Swe De Corp	Swedish International Development Corporation
T&D	Transmission and distribution
TDA	United States Trade and Development Agency
TOC	Total organic compounds
TSKJ	Technip, Snamprogett, M. W. Kellogg and JGC consortium
TVL	Threshold limit value
UNCTAD	United Nations Conference on Trade and Development
UNDP	United Nations Development Programme
UNEP	United Nations Environment Programme
USAID	United States Agency for International Development
USExim	United States Export-Import Bank
VOC	Volitile organic compounds
WASP	WIEN Automatic System Planning Package
WAPDA	Water and Power Development Authority (Pakistan)
YPF	Yacimientos Petroliferos Fiscales (Argentina)

Note: All dollar ($) figures are U.S. dollars unless otherwise indicated.

Conversion Factors

OIL

1 barrel = 42 US gallons = 35 Imperial gallons

1 long ton = 1.016 metric tons

1 ton crude oil = 7.33 barrels

1 ton crude oil = 40 MMBTU

1 million tons crude oil/year = 20,000 barrels/day

1 ton fuel oil = 6.7 barrels

GAS

1 cubic meter = 35.3 cubic feet

1,000 MMCFD = 10 BCM/yr

1 cubic meter NG = 36,000 BTU

1 cubic foot NG = 1,030 BTU

1,000 MMCFD NG = 7 million tons of LNG/yr

1 ton LNG = 15.5 barrels LNG = 9.53 barrels crude oil

1 ton LNG = 1380 cubic meters of NG

1 ton LNG = 52 MMBTU

ELECTRICITY

1 kWh = 3,412 BTU

1 BTU = 252 calories = 1,055 joules

1 thermie = 10^6 calories = 3,968.3 BTU

1 therm = 100,000 BTU

1 Gwh of electricity:

- Uses approximately 250 tons of oil in an oil-fired conventional steam power plant.
- Uses approximately 390-400 tons of coal in a steam coal-fired power plant.
- Uses approximately 8,000 MCF of NG in a combined-cycle power plant.

NOTES

tons or tonnes = metric tons

BTU = British Thermal Unit

MMBTU = million BTU

joule = a unit of energy (1 joule = 0.239 calories)

petajoules = 10^{15} joules

NG = natural gas

LNG = liquefied natural gas

MCF = 1,000 cubic feet

MMCFD = million cubic feet per day

BCM = billion cubic meters

kWh = kilowatt hour

Gwh = gigawatt hour (1 Gwh = 1,000,000 kWh)

PART I

THE SETTING

FINANCING ENERGY PROJECTS
IN EMERGING ECONOMIES

Part I consists of two brief chapters on the basic concepts and issues involved in financing energy projects in developing countries. It describes

- The concepts of public versus private financing, corporate-based (also known as *recourse*) and project-based (*nonrecourse*) borrowing, and general sources of equity and debt financing.

- The issues particular to funding projects located in developing countries, the implied project risks, and the objectives of risk mitigation.

1

Fundamentals of Project Financing

Before the 1970s, most petroleum projects in developing countries were financed by the internal cash generation of international oil companies (IOCs). During the 1970s, governments became heavily involved in the petroleum sector to ensure better control of their reserves and, in the case of petroleum-importing countries, to quell concerns regarding the security of oil supply. Consequently, oil and gas projects received increased financing from governments' budgets and official borrowings as well as from IOCs.

Since the early 1990s, most governments have limited their interventions in the sector and their budgetary contributions to it to encourage the private sector to undertake the required investments. The IOCs have also developed a tendency not to finance these projects from their internal cash resources. Their reasons have been twofold: first, low oil prices, shrinking margins, and large environmental costs have squeezed the IOCs' balance sheets; second, for a variety of reasons—including political need for local participation and desire to share project risks—IOCs have begun undertaking projects with a wide range of partners. As a result, funding of oil and gas projects has become quite complex, involving public and private investors and financiers.

Electricity supply in most developing countries has traditionally been undertaken by a vertically integrated public monopoly that generates, transmits, and distributes power to customer groups in various locations. Electricity supply is often viewed as a socioeconomic service and frequently involves general subsidies by the government, cross-subsidies among various classes of customers, or both. Required investments are generally funded from the utilities' internal resources and official borrowing from multilateral and bilateral agencies. In addition, in many instances utilities' funds are supplemented by government budgets.

The worldwide move toward further reliance on free-market systems is changing the economic structure of the power sector in many developing countries. In some countries, power sector assets are being privatized. In others, power utilities remain government corporations, but the private sector has been invited to build new power projects. Often, the private sector builds a power generating plant and sells its output to the public utility, which retains control of power transmission and distribution.

Transmission and distribution (T&D) networks are still viewed as natural monopolies for which government ownership or regulation is justified. Therefore, some countries have opted to maintain T&D facilities under public ownership, whereas others have aimed, at a minimum, to franchise distribution systems to private investors. For these reasons, new investors in power T&D are likely to be undertaken by both public and private corporations.

Overview 1.1 Public and Private Financing of Energy Projects

Energy projects, particularly gas and power facilities, will likely be subject to both public and private ownership in the foreseeable future. Accordingly, this book covers the sources and requirements for public (government or government-guaranteed) financing and private funding for oil, gas, and power projects.

RECOURSE AND NONRECOURSE FINANCING

Until the late 1980s, funds for long-term investments in the gas and power sectors in most developing countries were provided only by or through governments. A typical example would be investment in the construction of a new power plant by a public utility. The funds for such investment would come from (1) the internal cash generation (profit) of the public utility, (2) the government development budget, and (3) official borrowing from multilateral institutions (for example, the World Bank and regional development banks) and bilateral sources (for example, export-import banks). In such cases, the new project would be built as an extension of the assets of an existing public utility. Capital investment and borrowing would not be on the project account. Rather, the loans would be considered the public utility's debt, and as loan security the lenders would have full *recourse* to all assets and revenues of the public utility, not just those related to the new power plant.

The above arrangement is referred to as *corporate financing*. That is, borrowing is carried out by the public utility as a whole rather than by an entity created specifically to hold ownership of the new power plant. Although funds may be provided for investment in a specific project, lenders in a corporate financing arrangement look to the cash flow and assets of the entire company to service the debt and to provide security. Lenders thus carefully examine the company's financial track record, the viability of its expansion plans, and its projected financial performance.

An alternative to corporate financing is forming a *project company* specifically to construct the power plant. In this case, investments in construction of the power plant are viewed as assets of the project company. These funds come in the form of equity or debt. If the company is privately held, the equity is provided by the private owners. Additional funds are borrowed from a variety of sources. In such arrangements, the project's assets and cash flow secure the debt, and creditors do not have recourse to the sponsors' other available resources. Thus, borrowing for the project does not weaken a sponsor's general balance sheet or creditworthiness. Such self-standing borrowing, with no guarantees from sponsors (or governments) to lenders, is known as *nonrecourse* financing (it is also called *off-balance-sheet* financing).

In practice, however, most projects have *limited-recourse* financing, for which sponsors commit to providing contingent financial support (above their up-front equity commitment) to give lenders extra comfort. The extra security usually focuses on the construction and startup periods, which are the riskiest times for energy projects. For example, sponsors may agree to repay loans from sources other than the project company's assets (if project-company funds fall short) in cases of delays in project completion or cost overruns. Until the project passes final completion tests, then, lenders will have recourse to the sponsors' other assets and revenues. After completion, however, the lenders' right to repayment is limited to the income stream from the project and its underlying assets, including plant contracts and permits.

Overview 1.2 The Concepts of Recourse and Nonrecourse Financing

- Gas and power projects in developing countries are traditionally built as extensions of the assets of an existing company. For example, when a public utility constructs a power plant, funds are provided on the account of the entire company rather than on the account of the new plant. As loan security, lenders have full ***recourse*** to the assets and revenues of the entire company, rather than recourse only to those funds related to the new power plant.
- An alternative is the formation of a ***project company*** specifically for construction of the power plant. The project sponsors contribute equity. The assets and cash flow of the project itself secure debt, not the sponsors' other available resources. This type of borrowing, with no guarantees by sponsors (or governments), is described as ***nonrecourse***. Since the repayment of the loan is primarily dependent on the success of the project, lenders pay close attention to project risks.

MOBILIZING EQUITY AND DEBT FINANCE

Structuring a project-finance package entails deciding how much of the project resources should come in the form of equity; the rest will be project debt. Equity refers to funds put into the project company by shareholders of the company. Equity holders are owners of the company, and they receive dividends and capital gains based on net profits. Equity holders take risk; they would receive no dividend payments if the company lost money.

Project debt refers to funds lent to the project company by financiers such as commercial banks, insurance and pension funds, and multilateral institutions. These loans are secured by the project's underlying assets. Lenders receive payments for principal and interest on these loans whether the company makes or loses money. However, prospective lenders examine the company's projected cash flow very carefully to ensure that there is sufficient financial capacity for debt repayment.

A special type of funding called *quasi-equity* is occasionally used to attract risk-averse investors. In this case, a special class of company shares—preferred shares—is introduced. Payment to holders of preferred shares takes priority over dividends to ordinary shareholders but is subordinated to the claims of debt financiers and other creditors of the company.

Overview 1.3 Equity and Debt Financing

Funds to construct a new plant come in the form of either equity or debt.

- ***Equity*** is provided by owners, also called shareholders. Owners include project sponsors, who are the driving force behind a project, as well as "passive" investors, who hold equity positions without being involved in project promotion. Owners' return on their equity investment depends on the company's profit.
- ***Debt*** refers to funds lent by financiers. Lenders receive payments according to a predetermined rate (or formula). Their payment does not depend on the profit of the company.

For energy projects, equity varies between 20 and 40 percent of the project cost. Clearly, a higher equity ratio means a higher commitment by project sponsors and a lower risk for lenders. Thus, lenders like to see high equity ratios, whereas sponsors prefer lower ratios in order to minimize the funds they lock into one project. The acceptable equity ratio depends on the creditworthiness of the sponsors, the risks, and the location of the project.

Mobilizing equity used to be a straightforward process. A project sponsor would take funds from other resources and invest in the project company. Today, mobilizing equity involves much more innovative schemes. The possible sources of equity are

- Sponsor's own capital and subordinated loans, which are sometimes used to complement equity.
- Investment funds, a few of which have been recently formed to provide equity in private power projects.
- Multilateral institutions, such as the International Finance Corporation (IFC) and regional development banks, which have established new divisions for equity participation in private sector projects.

- International equity markets, where project sponsors can issue equity shares to the public or place shares privately with institutional investors.
- Local capital markets, where sponsors can issue equity shares to the public or to local institutional investors such as pension funds and insurance companies that may be interested in private equity placement in the project company.

Mobilizing equity becomes even more complex when a project is a joint venture between private and public corporations. In such a case, the equity share of the public corporation can come from the government and a host of sources of official lending agencies (the World Bank, regional banks, export-import banks, bilateral trade and aid agencies, and so on).

Debt financing requires a great deal of innovation. Indeed, during recent years debt financing has become more scarce than equity financing, at least for projects in developing countries. Traditional sources of debt financing—commercial banks—are not able to meet the financing needs of these projects. Furthermore, since the debt crisis of the 1980s, commercial banks have viewed large projects in developing countries as significantly risky, particularly if the loans are nonrecourse (that is, repayment is fully dependent on project revenues). Each bank has exposure limits to individual clients, sectors, and countries. Because a single bank can rarely meet all of the loan requirements of a major energy project, commercial banks now provide finance through syndications of lenders.

Commercial banks are also constrained by the time profile of their deposits. They cannot lend large volumes of long-term debt. Their loans offer maturities of 5 to 10 years, whereas most energy projects need financing for much longer periods. However, certain benchmark banks can encourage long-term participation of institutional providers in both debt and equity.

Faced with the limited availability of commercial bank loans, project sponsors utilize a variety of other methods to finance project debts, including

- Specialized energy funds sponsored by governments; for example, the Pakistan Energy Fund and the Jamaica Energy Fund, which provide debt financing to private sector projects.
- The IFC and most regional development banks, which offer loans to private companies without government guarantees.

- Suppliers' credit, which is extensively used to finance purchase of equipment and materials.
- International capital markets, which are increasingly tapped through issuing various types of bonds.
- Local capital markets in developing countries, which, although limited in scope, are being mobilized to finance project debt.

  In the case of joint ventures between private and public corporations, debt financing is also available through government-guaranteed official loans from multilateral institutions, regional banks, and bilateral agencies.

Overview 1.4 Sources of Equity and Debt Funds

Equity makes up 20 to 40 percent of project cost and is provided through

- Sponsors' own capital and subordinated loans.
- Multilateral institutions.
- International equity markets.
- Local capital markets.
- Certain investment funds.
- Governments, a host of official lending and aid agencies, or both, if the project is a joint venture between the private and public sectors.

Debt makes up 60 to 80 percent of the project cost and is provided through

- Institutional investors (pension funds, insurance companies, and mutual funds).
- International commercial banks.
- The International Finance Corporation (IFC) and regional development banks.
- International bond markets.
- Local banks and bond markets.
- Suppliers' credit.
- Specialized energy funds.
- Government-guaranteed official loans from multilateral institutions, regional banks, and bilateral agencies, if the project is a joint venture between the private and public sectors.

PROJECT RISKS

Essential to structuring a project finance package are identification, analysis, mitigation, and allocation of project risks. These risks are related to events that could endanger the project during development, construction, and operation.

At the stage of project development, the main risk is rejection by the government or financiers. Reasons for project rejections include commercial weakness and failure to obtain licenses, permission, and clearance. At the development stage, risks are high but involve relatively small losses, which are limited to money and time spent on feasibility studies and related preparatory work. These risks are borne primarily by project sponsors. However, in many energy projects, sponsors may be able to find multilateral and bilateral grants or other types of aid to finance part of the cost of project development.

During construction, the main risk is failure to complete the project with acceptable performance levels and within an acceptable time frame and budget. This risk falls mainly on the project company and its sponsors. They in turn hedge their risk by purchasing various forms of insurance and obtaining guarantees from contractors regarding costs, completion schedule, and operational performance of the project. The construction risk is high and potentially involves significant losses. This is a most important risk from the point of view of financiers. Should a project fail during construction, the lender's loan security—that is, the assets of the project company—would be of little value. Thus, financiers do not want to take any of the construction risk and normally ask for recourse to the sponsors' other resources until the project is completed and tested.

After the plant has been constructed, the main concern is that it may not operate on a continuing basis within acceptable economic and technical parameters. Such operational risks are numerous but are usually of more modest magnitude. They are related to technical failures, availability of fuels, market demand and prices, fiscal issues (taxes or subsidies), foreign exchange rates and convertibility, environmental problems, and so on. These operating risks are borne by the project company and its limited- or nonrecourse lenders. However, a project company can hedge against the risks through contractual and guarantee arrangements that in effect transfer some of them to other parties. The following are examples of such hedging:

- The project company receives guarantees from equipment suppliers for equipment performance.
- The project company obtains a supply guarantee from a fuel source, at defined prices that might be passed through or, in case of margin shortfall, with netback pricing down to a certain floor.
- The project company receives a take-or-pay contract from a company that buys project output at defined prices, sometimes including a pass-through of certain operating costs (for example, fuel costs).
- The project company receives guarantees against political risk from multilateral or bilateral agencies, channels foreign revenues through an offshore disbursement account, or both.
- The project company receives limited support from shareholders for defined margin shortfalls, through methods such as cash injection, subordinated loans, or dividend clawbacks.

The hedging arrangements affect not only the liability of the project company but also the willingness of private investors and financiers to support the project. This is the primary reason that project financing has turned into a complex discipline—financial engineering—involving a combination of instruments for guarantees, borrowing, and mobilization of equity.

Overview 1.5 Project Risks

- Essential to structuring a project finance package are identification, analysis, mitigation, and allocation of project risks.
- Project risks endanger the project during project preparation, plant construction, and plant operation.
- All risks are initially borne by project sponsors. However, project sponsors enter into numerous guarantee and contractual arrangements to hedge against risks.
- The practice of combining various instruments for guarantees, borrowing, and mobilization of equity is referred to as ***financial engineering*** and represents the heart of project finance.

2

Challenges of Project Financing in Developing Countries

Structuring financing packages can be especially difficult for projects in developing countries. The main reasons for this are as follows:

- Most developing countries have limited domestic resources. The main shortage is, of course, in capital funds. Other resources—such as skilled labor, raw materials, and infrastructure—may also be scarce.
- Because of the above shortages, prospective investors need to rely on foreign capital and other resources not locally available.
- In order to acquire foreign resources, investors must be convinced, and able to convince others (particularly financiers), that these resources will be utilized safely, efficiently, and profitably and that returns on investments can be repatriated in accordance with known rules and regulations.
- The above concerns are directly linked to the host country's business environment, including political and economic stability, legislative and regulatory systems, and labor disciplines.
- Imperfections in the country's business environment increase basic project risks and are magnified by a general reluctance to lend to the developing world.

Financing a project in the developing world is considered more risky than financing one in the developed world. Investors want higher returns to compensate for the higher risk. Investors are also forced to share risk by involving others in a project and to take measures to avoid or mitigate project risks. As a result, structuring a finance package involves many parties and instruments. This complexity is certain to increase with time.

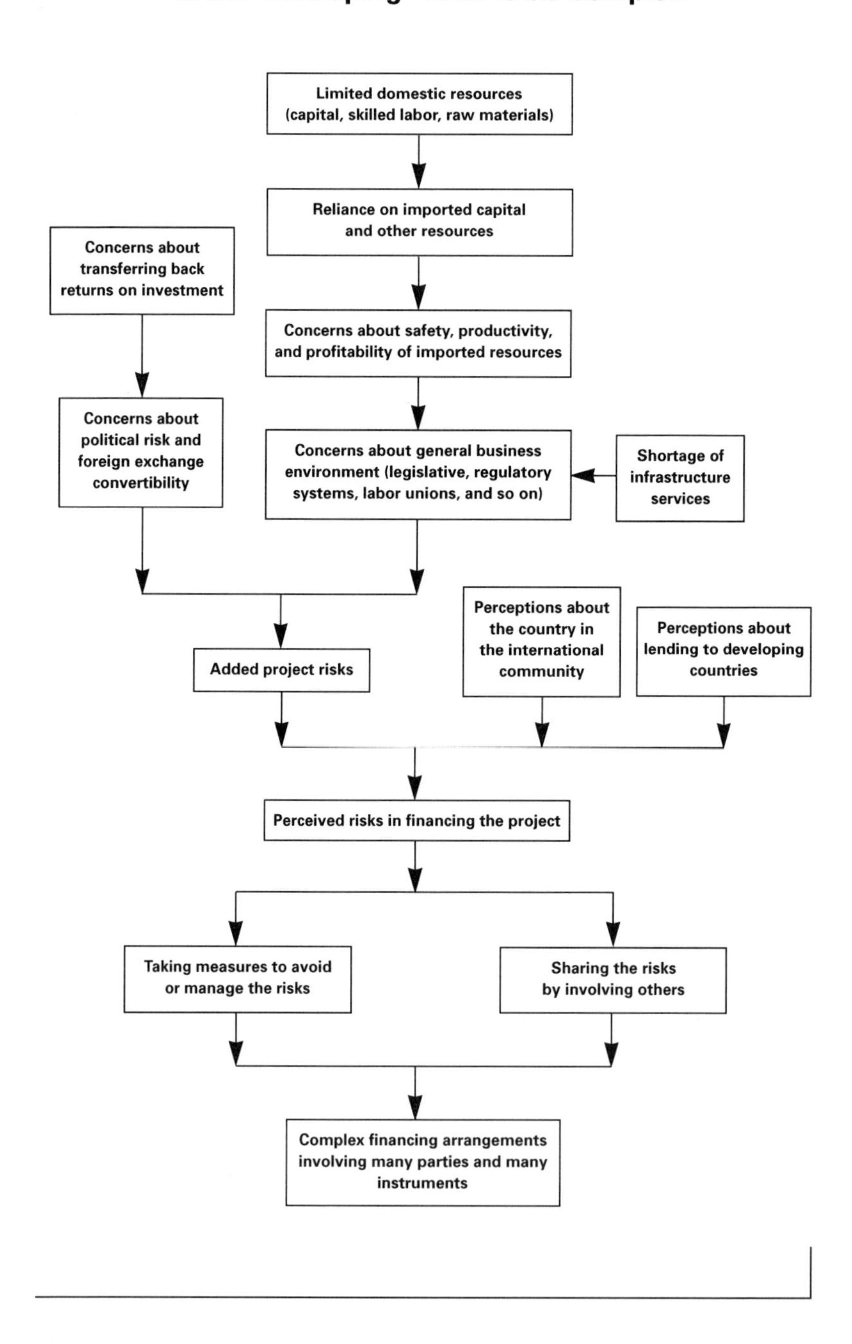
Figure 2.1 Why Financing a Project in the Developing World is So Complex
Limited domestic resources (capital, skilled labor, raw materials)
Reliance on imported capital and other resources
Concerns about transferring back returns on investment
Concerns about safety, productivity, and profitability of imported resources
Concerns about political risk and foreign exchange convertibility
Concerns about general business environment (legislative, regulatory systems, labor unions, and so on)
Shortage of infrastructure services
Perceptions about the country in the international community
Perceptions about lending to developing countries
Added project risks
Perceived risks in financing the project
Taking measures to avoid or manage the risks
Sharing the risks by involving others
Complex financing arrangements involving many parties and many instruments

THE DIFFERENCES IN BUSINESS ENVIRONMENTS

Business environments in developing countries differ from those of developed nations mainly in the following areas:

- Institutional and organizational structures and management.
- Legislative and regulatory systems.
- Economic and political security.

Institutional and organizational deficiencies in the energy sector normally relate to structural rigidities or vague divisions of responsibility. For example, the power industry in most developing countries has an integrated structure with a single national utility responsible for generation, transmission, and distribution. This structure is inherently inconsistent with a competitive environment. Although many of these countries try to create conditions that encourage private parties to enter the power sector, rigidities of the integrated system limit the potential for private participation. When private parties do formulate projects, they often find business arrangements difficult to conclude.

Organizational vagueness can delay projects and push costs above budgeted levels. Responsibilities of government agencies and national oil companies, with regard to petroleum development strategy, for example, often overlap. This divided responsibility results in delays in award of upstream contracts, construction of gas infrastructure systems, and so on.

Organizational deficiencies often originate in legislative and regulatory systems. However, these systems are also responsible for

- The lack of a transparent and open structure for dealing with private sector activities and, particularly, foreign investors.
- Inefficient systems of taxes and duties.
- Insufficient and inflexible power tariffs and oil and gas prices.
- Unclear safety and environmental standards.

Finally, economic and political insecurity are probably the most important reasons for concern about the business environment in developing countries. Economic insecurity affects demand and consumers' ability to pay for a project's output. Political insecurity can raise costs, reduce revenues, and in certain cases result in the total loss of investments or returns on investment. Risks resulting from political insecurity are as follows:

- **Expropriation.** This is the risk that property rights will be seized from the project by the host government, which often occurs without any compensation. Although fear of outright expropriation is always present, investors' greatest fear is creeping expropriation; that is, growth of the host government's take through increased taxes, royalties, fees, and so on, during the project's construction or operation.
- **Breach of undertakings by the host government.** This is the risk that the host government will not fulfill promises stated in the project agreement regarding contributions. These contributions can be financial ones, or they can be promises to build necessary infrastructure, such as roads, pipelines, or export terminals. The undertakings may also include such things as labor agreements and removal of certain regulations or governmental fees. Most host governments want to live up to their agreements because their reputations will be tarnished in the global investment community if they do not. However, changes in a country's political or economic landscape sometimes make it impossible to live up to commitments.
- **Civil unrest.** This is the risk that civil disturbances will cause disruption of the project's construction or operation. Disruptions can also physically damage the project. Obviously, the consequences to a project's financial return can be disastrous.
- **War.** The risk here extends to the total loss of lenders' and sponsors' capital. Even if complete loss does not occur, export lines could be disrupted, which would diminish the financial attractiveness of the project.
- **Expatriation of profits.** This is the risk that the host country will not allow profits realized from a project to be repatriated to the sponsors' home country. For example, the host country might insist that profits should be invested in other similar projects that may or may not be of financial benefit to the current project's sponsors.
- **Inconvertibility of developing country's currency.** This risk can have a substantial impact on a project that sells much of its product locally.

MITIGATING RISKS

International financiers, particularly commercial ones, have become quite cautious about lending funds to projects in developing countries. The debt crisis of the 1980s is still remembered by the banking and financial communities. Indeed, the banking system in the industrialized world has put strict regulations into effect for lending to developing countries. For example, in the United States, banks must adhere to larger reserve requirements when lending to countries that do not belong to the Organisation for Economic Co-operation and Development (OECD) than they do when lending to member countries; they are not allowed to take the front-end fees they usually earn as income at the time of the loan, as they can on OECD loans; and they must adjust their portfolios each year as part of their regulatory examination.

Overview 2.1 Major Concerns about Investing in the Developing World

- Financing a project in the developing world is considered more risky than in the developed world because of (1) deficiencies in institutional and organizational structures, (2) lack of clear and transparent legislative and regulatory systems, and (3) economic and political insecurity.
- These risks endanger the viability and sustainability of the project through (1) excessive construction and operation costs, (2) shortfall in revenue or the margin caused by price and market risks, and (3) uncertainty about safety and transferability of investments and returns.

Because of the above concerns and restrictions, banks seek a complete security package before they finance projects in developing countries. Ideally, the security package should protect the project against all significant risks. These risks are classified in a variety of ways but essentially relate to

- Costs (including construction and operations costs), which may be affected by inflation, interest-rate fluctuations, changing availability and rates of foreign exchange, delays, cost overruns, lack of raw materials or fuels, and so on.
- Revenues, which may be affected by price risk and demand risk, and combine with the above into *margin risk*.
- Safety and transferability of investments and returns.

Accordingly, measures to manage the risks try to convince financiers that

- Costs will not exceed the projected levels, and, if they do, some other party will take the burden before the cost increase affects the financing of the project.
- Revenues will not fall short of projected levels, and, if they do, some other party will make up the shortfall so that project finances are not hurt.
- Investment is safe and returns can be transferred out of the country, or, if funds cannot be transferred, a credible agency will cover legitimate losses.

PART II

GETTING TO KNOW THE FINANCIERS

Before attempting to structure the financing of a project, or even approaching financiers, project developers should become familiar with:

- The many sources of funds, including grants, soft loans, and technical assistance, that can be tapped for a project and are not normally known to project sponsors.
- The role that each financier normally wishes to play in a project. Most financiers would like to enter a project when the business environment is ready. Others, such as multilateral institutions, want to come in only if they can improve certain aspects of the business environment.
- The orientation of each financier. Some financiers like to support private sector participation, some want to promote certain technologies, some want to assist poverty reduction, and so on. Familiarity with these tendencies is required not only for choosing the right financier or financiers but also for packaging the project appropriately.
- The methods for combining the facilities provided by various financiers. Often this is necessary for energy projects because these projects are large and because many financiers prefer to join a project only when others are also participating.

Part II of the book familiarizes the reader with financiers and tells how to deal with them, in the following chapters:

- "Sources of Financing," which lists possibilities for financing oil, gas, and power projects.
- "Accessing Support from Global Multilateral Institutions," which explains the availability of assistance from agencies such as the World Bank and the International Finance Corporation.
- "Borrowing from Regional Development Banks," which describes the roles of Inter-American Development Bank, Asian Development Bank, and others.
- "Bilateral Sources of Financing," which covers the activities of development assistance programs and export credit agencies.
- "Accessing Commercial Funds," which explains the possibilities for mobilizing equity and debt financing from international and domestic capital markets. It also covers various types of guarantees and agreements that can be used to mitigate project risks.

3

Sources of Financing

Methods and sources for financing energy projects in developing countries have changed significantly over the past decade and a half. First, commercial banks have become more cautious about lending to developing countries after the debt crises of the 1980s. Second, a vast number of new instruments have been introduced to channel financial resources to equity and debt financing for energy investments. Third, energy corporations have developed a tendency not to keep project debt on their balance sheets for long periods. Fourth, governments increasingly prefer not to sponsor borrowing for oil, gas, and power projects. As a result of such developments, methods of financing have become more diverse. This change has taken a different course in oil projects than in power and gas projects.

FINANCING OIL PROJECTS

The financing of oil projects in developing countries has evolved through three distinct eras. Before the 1970s, most oil projects were financed by international oil companies (IOCs). Most IOCs had low debt-to-equity ratios; hence, they generally drew funds for exploration and development from internal cash flow and needed to do little borrowing. In addition, if the IOCs did have to borrow, they could raise long-term finance at favorable rates based on their creditworthiness and strong balance sheets.

During the pre-1970s era, the investment behavior of IOCs was determined more by strategic agendas than by strict financial criteria. In the 1970s and early 1980s, IOCs continued their active role in financing oil investments while governments, in both oil-exporting and oil-importing countries, began taking prominent roles in the petroleum sector. In oil-exporting countries, government involvement was motivated by the incentive to preserve their national resources and ensure the security of their share of

Figure 3.1 Changes in Methods of Financing Oil Projects

A: Financing Oil Projects Pre-1970

International Oil Companies

Commercial Banks

Corporate Financing

Internal Cash

Regional (Country) Company

Project 1

Project 2

Project 3

B: Financing Oil Projects in the 1970s and 1980s

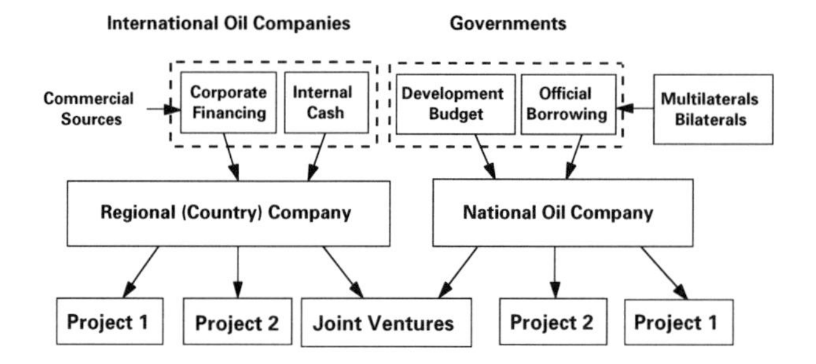

C: Financing Oil Projects in the 1990s

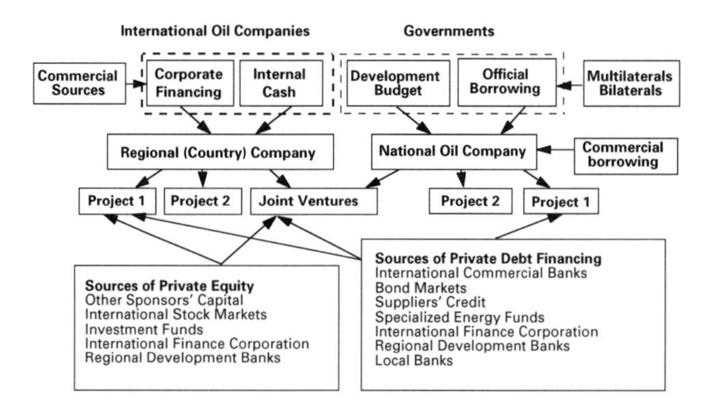

oil reserves. In oil-importing countries, interest stemmed from concern that security of oil supply was essential for national economic stability; thus, governments attempted to explore and develop domestic resources and to supervise petroleum imports firmly. Government intervention, enforced mainly through strengthened national oil companies, was accompanied by flows of state funds into petroleum projects or state-sponsored borrowing into equity and debt financing of such projects. In this period, IOCs, governments, multilaterals, bilaterals, and commercial banks were quite willing to invest in or finance oil projects.

During the second half of the 1980s and the early 1990s, however, almost all of the above sources of funds limited their exposure to oil investments. The most important reason was the collapse of oil prices in 1986 and the continuing softness of prices thereafter. The lower and uncertain petroleum prices not only reduced the attractiveness of oil investments but also imposed financial hardship on most IOCs. This hardship was exacerbated for some IOCs by new environmental standards for existing facilities, which translated into significant costs. As a result, IOCs restructured and adopted new, stringent investment standards. They have increasingly preferred shared risks and faster paybacks than they required in the earlier era. At the same time, the governments of both oil-exporting and oil-importing countries have substantially reduced their interventions and support of petroleum investments, for two primary reasons. First, domestic oil supply is no longer viewed as a national strategic matter. Second, there is a general tendency to limit the government role in activities that can be carried out by the private sector.

Although IOCs and governments still sponsor or provide financial resources for oil projects, a variety of new instruments are regularly employed to fund these projects. Private sources of equity and debt financing increasingly fund oil projects on nonrecourse and limited-recourse bases. Still, major oil companies prefer to fund projects from their own resources—including borrowing on their corporate accounts—rather than use project-based financing. For these companies, with their strong credit ratings, corporate debt is cheaper than project-based borrowing. However, many smaller IOCs are now involved in the petroleum sectors of developing countries, and these companies may not have very strong balance sheets and may not wish to borrow on their corporate accounts. In fact, even major oil companies borrow on a project basis when dealing with very large projects; when facing country-limit considerations (that is, when a company's exposure in one country is too high or higher than the company's strategic allowance); or when they are involved with partners that cannot afford to borrow on the strength of their own balance sheets.

Overview 3.1 Methods of Financing Oil Projects

- In the 1990s, international oil companies (IOCs) have become more financially stringent in regard to their investment decisions in developing countries. Furthermore, the governments of these countries have reduced their financial support for petroleum development and supply.
- As a result, IOCs have become much more selective in undertaking investments and have sought to share project risks by involving other parties, particularly local partners. These preferences have led to more complex financing arrangements addressing the constraints of country limits or weaker (or less creditworthy) partners. Straightforward corporate financing is no longer fully applicable. Rather, various sources of funds and guarantee instruments are needed.

FINANCING POWER PROJECTS

Before the 1950s, power supply in most developing countries was run by small private companies that served specific cities and locations. In the 1950s and 1960s, the governments of these countries undertook extensive electrification programs that aimed at providing electricity to all social and economic groups. Along with the implementation of these programs, power grids were interconnected (for the most part) to take advantage of economies of scale in power generation. Ownership of power supply facilities was also consolidated into national (or state) utilities. These state-owned utilities, which were directly or indirectly run by their corresponding governments, became responsible for power generation, transmission, and distribution. Power supply came to be considered a socioeconomic and strategic matter with natural monopoly characteristics. Private sector participation was not deemed appropriate. Indeed, most countries had legal barriers to private sector entry into the power sector.

Figure 3.2 Changes in Methods of Financing Power Projects

A: Traditional Power Project Financing

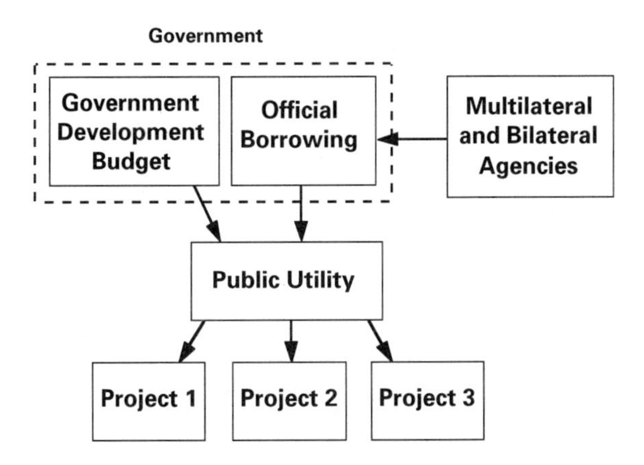

B: Power Project Financing in the 1990s

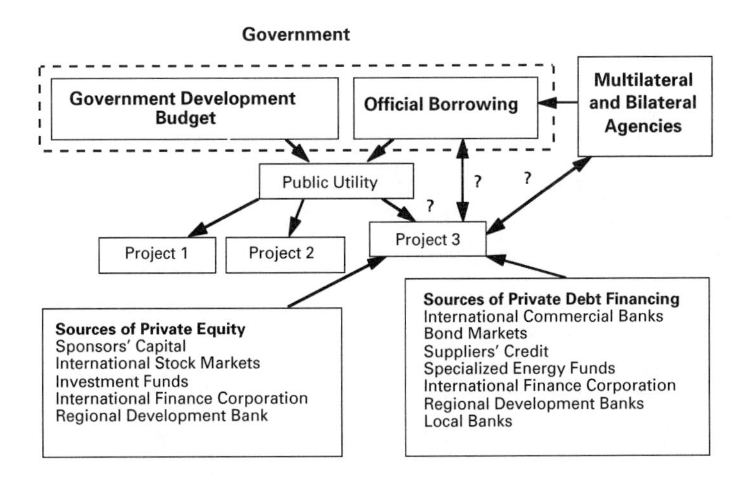

During this period, power project financing in developing countries involved public utilities raising funds from the government development budget or through official borrowing. Official borrowing was financed by multilateral and bilateral agencies and some commercial sources. Generally, the public utility would borrow funds, with government backing, usually in the form of a government counterguarantee.

Overview 3.2 Methods of Financing Power Projects

- For power projects, the traditional source of financing has been the government budget or government-sponsored borrowing.
- The inability or unwillingness of governments to continue supporting power sector investments has (1) allowed and facilitated the entry of private sector investment and (2) encouraged public utilities to borrow from commercial sources and even to raise equity on international and domestic capital markets.
- Private investment in the power sector is concentrated in generation. The emerging trends toward corporatization and privatization of power systems are expected to result in some private investor participation in transmission and, even more so, in distribution. However, financing for these facilities is expected to remain corporate-based, because most new investments are aimed at expanding existing networks and are financially intertwined with previously built assets.

In the early 1990s, the governments of most developing countries decided to limit their involvement and obligations to fund the power sector by allowing the private sector to undertake part of the required investments. Governments also encouraged state public utilities to raise private equity and finance. As a result, in the first half of the 1990s, a number of power plants were financed with private funds on a project basis, although many power plants were still built by state utilities. Private financing of power transmission and distribution is limited to exceptional cases, where these systems have been privatized. Such projects are financed mostly on a corporate basis because most new projects are expansions of existing networks and are intertwined with existing facilities.

The methods of financing power investments are expected to change further as developing countries corporatize and eventually privatize their power systems. Emerging methods of financing may involve using project-based financing for new power plants and corporate-based financing for transmission and distribution projects.

FINANCING NATURAL GAS PROJECTS

Financing of natural gas projects is similar in some ways to financing of oil and in other ways to that of power projects. Natural gas transmission and distribution in most developing countries have been viewed as constituting a natural monopoly and treated like the power sector. Hence, gas transmission and distribution investments have historically been undertaken by state entities. However, upstream developments have been modeled after those of the oil sector. It has therefore been acceptable to leave upstream gas development to private companies or to joint ventures between the public and private sectors.

Factors similar to those that influence the power sector have caused many developing countries to plan to corporatize and eventually privatize their gas utilities. After privatization, investment requirements for gas systems will logically have to be funded by private sources. In certain cases, for example, when a main transmission pipeline or a new distribution network needs to be built, project-based funding will likely be possible. However, in other cases, when investments are needed to expand existing networks, financing will probably have to remain corporate-based.

Overview 3.3 Methods of Financing Gas Projects

- Financing of gas projects has some similarities with that of both oil and power. Gas transmission and distribution are viewed together as a natural monopoly and treated like the power sector. Upstream development is modeled after the oil sector and mostly left to private companies or joint ventures between private and public entities.
- In the event that gas transmission and distribution systems are privatized, new investments will likely be funded by private sources.

SOURCES OF FUNDS

Analysis of the methods of financing oil, gas, and power projects indicates that

- The role of the public sector will remain important into the foreseeable future. Therefore, public and private investments will complement each other for most energy projects.
- Public (that is, state-sponsored) projects will not be purely public because most state entities will be encouraged to borrow from or even to mobilize equity from commercial sources.
- Private-sector-sponsored projects will not be purely private because most sponsors will receive assistance from governments and from official development agencies in securing debt financing and even in mobilizing equity.
- Project-based financing will not be purely on a nonrecourse basis because sponsors will likely continue to find out that some type of corporate or even government guarantee can help in raising equity and debt finance at substantially lower costs.

For the above reasons, and because of the large size of most energy projects, a project planner should at least consider all types of funds, whether the project sponsor is a state entity, a project company, or a well-established corporation. These sources of financing include

- Multilateral development institutions, which were traditionally supporters of state entities but have now introduced a variety of facilities to assist private sector investment and finance. Multilaterals include global institutions, such as the World Bank and the International Finance Corporation (IFC), as well as several regional development institutions that have the same objectives as the World Bank but concentrate on specific regions. The contribution of multilaterals to investment in and financing of energy projects does not exceed $10 billion annually, compared with overall annual investment requirements of more than $150 billion. Despite their relatively small contribution to the total financial requirements of the energy sector, the multilaterals' participation in any project is considered quite important because of the comfort it provides to other participating investors and financiers.
- Bilateral agencies, such as the export-import banks of industrialized countries, which traditionally lent to the state entities of developing countries to promote supply of locally produced equipment, are now

Overview 3.4 Sources of Financing Energy Projects

Because of market circumstances and the large size of most energy projects, a project sponsor—whether a state entity, a well-established corporation, or a new project company—must consider all sources of financing, including

- Multilateral development institutions.
- Bilateral agencies, such as the export-import banks of industrialized countries.
- Commercial banks and institutional providers, equity markets, bonds, and specialized energy funds.
- Ad hoc sources, such as equipment suppliers' credit, project contractors' financial contributions, and equity and debt financing by purchasers of the project output (public utilities, trading houses, and so on).

willing to lend to both state and private entities under certain conditions. Bilateral agencies provide finance in the forms of suppliers' credit, buyers' credit, and guarantees. However, their assistance is often tied to their own national interests.

- Commercial finance, which encompasses many sources and instruments. Commercial sources of equity include stock markets and specialized energy funds. Sources of debt financing include commercial banks and long-term institutional providers (of debt and equity); bond markets; and, again, specialized energy funds. State and private entities use commercial financing. Providers of these funds normally require a demonstration that all project risks have been identified and managed effectively, likely necessitating a host of guarantee instruments.
- Ad hoc sources of financing, which depend on the project and country. For example, governments of developing countries are still important sources of support for energy projects, particularly gas infrastructure and power systems. Governments' roles as direct financial contributors will likely continue to decline. Governments may, however, continue to provide subsidized energy, tax holidays, and exemptions from custom duties. Other ad hoc sources of finance include (1) project contractors, who are sometimes willing to contribute to investments and finance in order to win business; (2) sources of equipment, which are usually large

multinational companies with substantial resources; and (3) purchasers of output. With regard to the third point, public utility corporations (as purchasers of gas and power) or large trading houses (as purchasers of oil) may be willing to make contributions to equity and debt financing of energy projects.

4

Accessing Support from Global Multilateral Institutions

The origins of international development institutions go back to a gathering of delegates from 44 countries, who met in 1944 to design a framework for economic cooperation following World War II. The gathering, known as the United Nations Monetary and Financial Conference, was held in Bretton Woods, New Hampshire, and resulted in the creation of two institutions: the International Monetary Fund (IMF) and the International Bank for Reconstruction and Development (IBRD), which is also known as the World Bank. The IMF was created to monitor the macroeconomic and monetary policies of member countries to ensure that the misguided, nationalistic monetary policies of the 1930s did not reappear in the postwar period. The World Bank was initially designed to provide financing to rebuild roads, communication links, power systems, and other basic building blocks of the economies of war-torn Europe.

Comprehending the roles and involvement of the IMF and the World Bank is essential for project financing in developing countries because

- Both institutions have great influence on the business environments of developing countries.
- Direct financial support from these institutions significantly affects the availability of funds from other sources for financing energy projects.
- The World Bank and its affiliates provide direct support in the form of equity, loans, and guarantees for financing oil, gas, and power projects.

The IMF and the World Bank are nonprofit international organizations that belong to the governments of the world. As noted, the original Bretton Woods conference was attended by representatives of 44 countries, the largest of which were the United States, the United Kingdom, France,

China, India, and the Soviet Union. The nations attending the conference, except the Soviet Union, joined the membership of the IMF and World Bank. Over the years, many other countries joined, and in 1991-92, 15 republics of the former Soviet Union became members of the two institutions. The total membership in 1995 was 179 countries.

Overview 4.1 Origins and Significance of International Development Institutions

- The origins of the major international development institutions go back to the Bretton Woods conference in 1944. The conference resulted in creation of the International Monetary Fund (IMF) and the International Bank for Reconstruction and Development (IBRD), also known as the World Bank.
- Comprehending the roles of the IMF and IBRD is essential for project financing because both institutions have great influence on a project's business environment. Their financial assistance significantly affects availability of funds from other sources. In addition, the IBRD and its affiliates provide direct support for financing oil, gas, and power projects.
- The IMF and IBRD are owned by the governments of their 179 member countries (1995). The share of each country is determined based on the size of its economy.

The organizations are owned by their member countries. The share of each country is determined by a formula based mainly on the economic capacity of the country compared with those of the other members. The 10 largest shareholders of the World Bank are the United States (17.48 percent), Japan (6.41 percent), Germany (4.95) percent), France (4.75 percent), the United Kingdom (4.75 percent), Canada (3.06 percent), China (3.06 percent), India (3.06 percent), Italy (3.06 percent), and Russia (3.06 percent). A country holds voting power on the board of each institution in proportion to its shareholding.

The IMF and the World Bank are referred to as sister organizations. Their headquarters are located in Washington, D.C. The World Bank now has three affiliates: the International Development Association (IDA), the International Finance Corporation (IFC), and the Multilateral Investment Guarantee Agency (MIGA). In addition, the World Bank houses several other important programs, such as the Global Environment Facility (GEF) and the Energy Sector Management Assistance Programme (ESMAP).

THE INTERNATIONAL MONETARY FUND

Although the IMF does not have direct involvement in project finance, its role, its influence, and particularly its position vis-à-vis governments of developing countries affect the challenge of bringing potential financiers into energy projects.

At the time of its creation, the IMF's function was to maintain the par-value system of fixed but adjustable exchange rates. It was intended to oversee the economic policies of its members and to use its financial resources to help them through periods of adjustment when extreme balance-of-payments difficulties arose. It played this role for two and a half decades, during which its jurisdiction included both industrialized and developing countries.

In the 1970s, however, major industrial countries abandoned the par-value system, and the IMF's core mission of stabilizing the international monetary system thus disappeared. Subsequently, the IMF adjusted its function and became an important player in recycling petrodollars in the 1970s, managing the debt crises in the 1980s, and transforming former communist countries in the 1990s.

The IMF operates in two distinct ways. First, it tracks its members' monetary and fiscal policies, which can influence governments' ability to finance imports and exports. During regular consultations with government officials, the IMF recommends policy changes to correct current problems or head off future ones. Second, the IMF provides loans to member countries (rich or poor) that have short-term problems in meeting foreign payment requirements. These loans are normally based on tough policy conditions that call for establishment of sound fiscal and monetary parameters.

Particularly in the granting of loans, the IMF's policy interventions can extend to tax, tariff, and pricing issues, which are very important from the point of view of financing energy projects. For example, in developing countries where energy prices are kept artificially low, the IMF normally pressures governments to move toward cost-based and market-related pricing of power, petroleum products, and natural gas. Such moves usually increase the commercial viability of projects in these sectors. The IMF further influences project financing when its position conflicts with government fiscal and monetary policies. When this happens, the country's credibility suffers significantly, which adds to the challenge of financing energy projects within its borders.

Overview 4.2 The International Monetary Fund

- The IMF oversees the economic policies of its member countries (rich and poor) and uses its financial resources to help them through periods of adjustment.
- During its regular consultations with government officials, the IMF encourages them to take remedial action to correct problems such as inflation, unemployment, and balance-of-payments deficit.
- When providing a loan, the IMF pushes for tough policy actions for establishing sound fiscal and monetary parameters.
- The IMF's policy interventions extend to taxes, tariffs, and energy pricing, which are important from the point of view of financing oil, gas, and power projects. The IMF's overall attitude toward a country affects the perception of the country's risks.

INTERNATIONAL BANK FOR RECONSTRUCTION AND DEVELOPMENT

The IBRD is the original and main component of the World Bank Group (IBRD and *World Bank* are used interchangeably). The term *World Bank Group* refers to the IBRD and its affiliates: the IDA, IFC, and MIGA. When the IBRD was founded in 1944, it operated on the principle that European countries would be short of foreign exchange for reconstruction yet would be insufficiently creditworthy to borrow commercially. As an official multilateral institution whose share capital was owned by countries in proportion to the size of their economies, the World Bank would serve as an intermediary by borrowing in world capital markets and lending to member countries in need of foreign capital. The first loans awarded by the World Bank helped finance reconstruction of the war-ravaged economies of Western Europe. However, very soon the mission of the World Bank was extended to economic development (in addition to reconstruction) of member countries.

The objectives of the World Bank are to help reduce poverty and to foster economic development. Although these objectives have remained the same, the philosophy and composition of the World Bank's activities have changed markedly over time. For two to three decades, the World Bank's main activity was financing investments in infrastructure, social programs, and economic activities. The decision to finance depended on the soundness of a project and its return compared with alternatives and cost of finance. Projects included roads, power plants, schools, and irrigation networks, which were viewed as essential for economic and social development.

Over the years, the composition of the World Bank's activities has changed in response to changes in the world economic and political climate, the influence of stronger shareholders, and the learning process involved in development of member countries.

In the 1960s, the World Bank based its activities on the conviction that providing financing does not, by itself, lead to success if the institutions responsible for implementation and operation of a project are not competent. Accordingly, most projects since then have included a component for institution building and management training. Many of today's strong power utilities were created or reinforced through institution-strengthening components of World Bank loans.

In the 1970s, the conviction was that a project could not succeed unless it was operated not only by a competent institution but also in the right economic environment. Attention was then extended to correcting pricing and other policy matters affecting the business environment. This meant that a project financed by the World Bank included components for institution building and conditions or action plans to adjust prices, taxes, subsidies, and so on. In the meantime, external shocks resulting from the oil crises of the 1970s added substantial instability to the business environment. The World Bank was then drawn into fixing the economic environment not only through its project lending but also through structural adjustment loans. These latter loans provided a country with balance-of-payment support; that is, foreign exchange that could be used by a government for almost any need (excluding military spending). In return, the government was required to commit to a specified list of actions (for example, adjusting exchange rates, removing subsidies, and so on) considered necessary for creating a healthy economic environment.

Overview 4.3 The Objectives of the World Bank

- The International Bank for Reconstruction and Development (IBRD) is also referred to as the World Bank. However, the term ***World Bank Group*** refers to the IBRD and its affiliates: the International Development Association (IDA), International Finance Corporation (IFC), and Multilateral Investment Guarantee Agency (MIGA).
- The World Bank's mission was initially to reconstruct Europe after World War II, but this was extended to cover economic and social development throughout the world.
- The World Bank's objectives of reducing poverty and stimulating economic growth have remained the same, but its philosophy and the composition of its activities have changed markedly over time.
- In its early years, the World Bank financed projects that were sound and useful for economic and social development. Today, a host of additional considerations, including pricing issues, institutional and regulatory matters, encouragement of private sector participation, and mitigation of environmental risks, come into play when a loan is granted.

In the second half of the 1980s and the first half of the 1990s, the World Bank shifted its emphasis to reducing the role of the public sector and encouraging moves toward a free-market system. The World Bank's objectives are still to reduce poverty and to foster economic development. However, it now tries to achieve these objectives by creating the right business environment, in which the private sector can take the lead role in formulating, implementing, and operating most projects. In addition, the World Bank has become quite conscious of the environmental impacts of development projects. Thus, to make a project attractive to the World Bank, a project sponsor has to demonstrate how the project and the involvement of the World Bank would improve the business environment, facilitate private sector participation, or ameliorate environmental problems. Because World Bank officials consider their agency the lending source of last resort, demonstration that a project would not materialize without assistance from the World Bank is also helpful in acquiring funding.

The World Bank annually lends $17 to $20 billion, of which $2 to $3 billion is allocated to power projects and about $1 billion to oil and gas projects. The World Bank's involvement in the power sector has remained relatively stable over time. However, its lending to the oil and gas sector has fluctuated substantially, primarily because of external factors. Thus, lending to the oil and gas sector became substantial in the late 1970s and the early 1980s, when security of petroleum supply had become an international concern. The World Bank's objective was to help developing countries explore and develop indigenous petroleum resources. By 1983, its lending to the oil and gas sector had reached $1 billion per year. However, private oil companies perceived that the World Bank was preempting activities of the private sector and exerted pressure on the World Bank to limit its activities in the sector. As a result of this controversy and of the collapse of energy prices in 1986, World Bank lending to the oil and gas sector dropped sharply in the second half of the 1980s. By the first half of the 1990s, however, the World Bank had become quite active in the oil and gas sector as a facilitator of private sector participation. Furthermore, World Bank activity had become helpful in natural gas development, which had emerged as a major area of investment and finance in many client countries.

The World Bank needs a governmental guarantee of repayment of its loans. Therefore, the World Bank's financial assistance is not directly extended to private financing of a project. However, World Bank involvement in a project helps and encourages private sector participation in projects in many ways:

- In joint ventures between the private and public sectors, the World Bank finances the equity or debt share of the public sector in the project.
- World Bank participation in a project often creates comfort for other financiers and investors to enter into the project.
- The World Bank can extend guarantees for political risk, foreign exchange convertibility, and so on.
- The World Bank's influence on or closeness to the host government helps facilitate legal and administrative preparation of the project package.

Overview 4.4 The World Bank's Lending to the Energy Sector

- The World Bank annually lends $17 to $20 billion, of which $2 to $3 billion is allocated to power and $1 to $1.5 billion to oil and gas.
- The World Bank requires a government guarantee for repayment of its loans. Thus, its loans are normally made to governments and government affiliates.
- The World Bank assists private sector financing of projects through financing of public sector shares in joint ventures, through mobilizing other financiers, and through its guarantee instruments. In certain cases, arrangements have been made for World Bank loans to support private investments through government-affiliated intermediaries.

How a Loan Is Processed

An overview of the World Bank's loan-processing procedure is useful because it shows how the World Bank interacts with its borrowing countries and because the World Bank's procedures are very similar to those of other multilateral institutions.

The World Bank has a formal loan application process that includes project identification, project preparation, project appraisal, loan negotiations, and a presentation to the World Bank's Board of Executive Directors. The first phase, identification, is clearly essential for the project to be brought into the lending pipeline. Although only governments can propose projects for financial assistance, identification can come from several sources, including background work of World Bank staff, of other United Nations agencies, or of private sponsors. In the energy sector, many projects are identified during the occasional studies of sectoral policy and investment strategies that the World Bank conducts for its client countries. ESMAP, a joint United Nations/World Bank effort, often reviews energy sector issues and identifies energy projects for financing by the World Bank and other financiers. At the identification stage, details of the project are not known, but the project must appeal to World Bank objectives in order to be processed.

Project preparation begins once the World Bank and a government agree on objectives. The borrower is normally a public company—for example, the state power or gas utility or the national oil company. The government is the guarantor but also the decisionmaker about how much the country should borrow and how the borrowed funds should be allocated to different sectors. Preparation is the responsibility of the borrower, although the World Bank often provides substantial assistance or at least guidance. Project preparation entails development of an idea into a detailed proposal that considers all aspects of the project—technical, economic, financial, social, institutional, and environmental. Most borrowing entities recruit international consulting firms to assist them in project preparation. At this stage, the presentation and packaging of the project are very important. The technical and economic soundness of the proposal should be firmly demonstrated by comparing the project with its alternatives. Other external benefits should also be explained and shown to satisfy the major objectives of the World Bank.

After preparatory work has been completed, a project is considered ready for appraisal. The appraisal is viewed as an independent assessment of the project by the staff of the World Bank. Its assessment includes technical, economic, financial, institutional, and environmental considerations. If project preparation is done within the framework discussed above, the appraisal process is likely to go smoothly. Questions that an appraisal mission seeks to address are usually the same as items described above. Results of the appraisal work are included in a Staff Appraisal Report for the project. This report is used to prepare the legal documents—normally a Project Agreement—between the World Bank and the borrowing entity and a Guarantee Agreement between the Bank and the government.

The Appraisal Report and legal documents are discussed in a negotiation session between the World Bank staff, representatives of the borrowing entity, and the government. Through a give-and-take process, the World Bank and the borrower review all the issues that have arisen during preparation and appraisal. They then agree on an action plan to resolve the issues. For example, if electricity prices in the country are too low for commercial viability, the World Bank and the borrower would agree on an action plan for a gradual increase in power prices. The negotiation agreements normally include covenants to improve the institutional and regulatory framework of the energy sector, to move toward market operations, and to improve environmental and safety aspects of the energy sector. Many of these covenants go beyond the formal scope of the project; this is what "development impact" is about from the point of view of the World Bank.

Overview 4.5 How to Make a Project Attractive to the World Bank

- To make a project attractive to the World Bank, the sponsor should demonstrate that the project and involvement in it by the World Bank would improve the business environment or facilitate private sector involvement or environmental remediation. It is also useful to show that the project would not materialize without the World Bank's involvement.
- Preparing an attractive project package should include demonstrating its technical and environmental soundness, economic desirability, and safe design, as well as the institutional and financial competence of the borrower.
- Many of the World Bank's loan conditions go beyond the formal scope of the project; this is what "development impact" is about, at least from the point of view of the World Bank.

The final stage of loan processing is the presentation to the Board of Executive Directors. The Appraisal Report and the legal documents are reviewed by board members for approval. Board members function in continuous session at the World Bank's headquarters in Washington, D.C., and meet as often as the business of the World Bank requires. The board consists of 24 executive directors. The five largest shareholders—the United States, Japan, Germany, France, and the United Kingdom—each appoint one executive director. The other countries are grouped in 19 constituencies, each represented by an executive director elected by a country or a group of countries. The number of countries represented by the 19 directors varies widely. For example, the executive directors for China, Russia, and Saudi Arabia represent one country each, whereas a single director speaks for 24 Francophone African countries. The country groups are formed more or less along geographic lines. However, members themselves decide how they will be grouped, and each director carries a voting power equal to the sum of the capital shares of the countries in the corresponding group.

Overview 4.6 How the World Bank Processes Loans

- The World Bank project cycle includes project identification, project preparation, project appraisal, loan negotiations, and board presentation.
- The Board of Executive Directors of the World Bank consists of 24 executive directors who represent the governments of the 179 member countries.
- Board approval of a loan is based not only on the merits of the project but also on the strategic and political agendas of board members.

Board approval of a loan is based on the merits of the project and the strategic and political considerations of board members. Normally, by the time a project reaches the board session, the expected concerns have already been addressed, and board approval is quite likely.

The World Bank's Guarantee Instrument

Since the debt crisis of the 1980s and the ensuing retreat of commercial banks from financing of projects in developing countries, the World Bank has been experimenting with a number of guarantee instruments to encourage lending by commercial financiers. Until the early 1990s, these experiments were limited in scope and success and isolated from the main core of the institution's operation. In 1994, however, the World Bank decided to "mainstream" its guarantee operations.

Within the new framework, guarantees are issued to private lenders to cover risks arising from nonperformance of sovereign contractual obligations or from *force majeure* aspects of a project. This instrument is called a *partial guarantee*, denoting that private lenders still bear other project risks.

The World Bank partial guarantee instrument is especially helpful for private sector projects, where the financial viability of the project hinges on contractual arrangements with a number of state entities. The instrument can guarantee against defaults in sovereign contractual obligations, such as

- Government adherence to an agreed tariff formula.
- Delivery of inputs, such as fuel supplied to a private power plant.

- Payment for outputs, such as power supplied by an independent power plant to a state public utility.
- Compensation for project delays or interruptions caused by government actions or political events.

In addition, transfer risks for investors and lenders may arise because of constraints in availability of foreign exchange. The World Bank guarantee can cover the risks of converting foreign exchange.

The World Bank's partial risk guarantee is triggered when a government does not comply with one or more of the obligations stipulated in agreements with project sponsors, which constitutes a default in debt service to lenders.

The World Bank guarantee instrument is designed to help developing countries gain access to a wider range of private lenders and to borrow at lower costs.

Unlike MIGA, the World Bank does not guarantee equity capital. The World Bank also does not guarantee loans from other official multilateral financial institutions or export credit agencies. The World Bank guarantee requires a counterguarantee from the government. Under a partial guarantee, up to 100 percent of the principal and interest can be covered. The World Bank charges two fees for the guarantee cover: a *standby fee* and a *guarantee fee*. These fees are charged either to the borrower or to the lender. The fees are determined on a case-by-case basis.

Where Does the World Bank Obtain its Funds?

Like any financial institution, the World Bank has its own capital base, which comes from its member governments. However, when member governments buy shares they pay in only a small portion of the value of each share. The unpaid balance is on call should the World Bank be unable to pay its creditors—something that has never happened. Thus, the paid-in capital is not significant compared with the World Bank's outstanding loans of $50 billion. The World Bank raises most of its money on the world's financial markets. It sells bonds and other debt securities to pension funds, insurance companies, corporations, other banks, and individuals around the world. Because of the backing of almost all the governments of the world and cautious financial management, World Bank bonds have the highest rating, which enables the World Bank to borrow on good terms. This advantageous position and the World Bank's non-profit status make its loan rates better than those on commercial loans. World Bank loans, however, are not soft, subsidized, or cheap. There is widespread public confusion about the differences between loans from the World Bank and loans from its affiliate, the IDA. IDA loans are soft. However, these are part of a wholly different process, which is explained in the next section.

THE INTERNATIONAL DEVELOPMENT ASSOCIATION

IDA was established in 1960 as an affiliate of the World Bank to provide assistance to the world's poorest countries. It provides interest-free loans—called credits—to countries with annual per capita income of less than $800. IDA has 155 member countries. Many rich member countries (such as the United States, Japan, Germany, and the United Kingdom) and middle-income members (such as Brazil, Hungary, Korea, and Mexico) make financial contributions, which are made available to poor members. Of the 155 member countries, 77 are eligible to borrow. Some 21 countries that were once borrowers have "graduated" from IDA, and two—Korea and Turkey—are now among the donor members. IDA's resources are replenished every three years. Originally, grants were designed so that each donor's contribution was proportional to its capital in the World Bank. Subsequently, donor contributions to IDA replenishments have been determined through negotiations taking into account a number of economic criteria such as GNP, per capita income, and trade-related indicators.

As noted, there is widespread public confusion about the distinctions between IDA and the World Bank. IDA is, in many respects, indistinguishable from the World Bank. Both institutions finance development projects and aim at reducing poverty. IDA has the same staff as the World Bank, and the president of the World Bank is also the president of IDA. However, each institution has separate Articles of Agreement, different provisions for paying in capital subscriptions, different voting structures, and separate financial resources.

The fundamental difference between the two institutions is the way they obtain funds and the terms on which they lend to developing countries. The World Bank raises most of its funds on the world's financial markets and lends to developing countries at interest rates somewhat below those of commercial banks. By contrast, IDA provides the world's poorest countries with interest-free credits. Because of its highly concessional loan terms, IDA cannot raise funds in capital markets. Instead, its resources come from contributions by donor governments.

IDA's central objective is poverty reduction, to which it takes a twofold approach:

- Supporting a pattern of economic growth that will provide efficient employment and income opportunities for the poor while avoiding subsidies for capital-intensive industries or damaging the environment.
- Supporting investment in people in primary health, education, and nutrition with the objective of enabling the poor to participate in economic growth.

Based on this approach, IDA provides credits to countries that are in serious poverty but show a determination to take the policy actions required for sustainable development.

IDA's annual lending is $5 to $7 billion. Most IDA credits are extended for specific investments, although about one-quarter of lending supports structural adjustment programs. Because of its emphasis on reducing poverty, IDA pays special attention to social sectors such as education, health, and nutrition. However, infrastructure projects, among them electricity supply systems, receive substantial portions of IDA credits. On average, 25 percent of IDA resources are allocated to financing infrastructure projects.

Like the World Bank, IDA has changed its philosophy of lending over time. Although at the beginning IDA saw its role as the financing of economically sound projects, it now looks for wider economic and social impact from its financing efforts. For example, in the power sector, the emphasis has shifted from building new projects to rehabilitating existing ones, along with improving pricing policy, regulatory processes, and institutional structures. IDA also has developed an emphasis on expanding the role of the private sector and dealing with environmental concerns.

IDA lends to governments or government-affiliated entities. Its loans could finance equity or debt shares of a public entity in joint venture with private companies. The project cycle and preparation arrangements are similar to those of the World Bank. However, because of its favorable loan terms, IDA has greater leverage vis-à-vis its borrowers for improving the project's business environment or the energy sector as a whole.

To be of interest to IDA, a project, and the involvement of IDA in it, must result in noticeable reduction in poverty. Other developmental impacts, such as facilitating private sector participation and reducing environmental damage, are also important.

In addition to providing direct support to poor countries, IDA plays an important role in mobilizing and coordinating aid from other multilaterals and from individual donor countries. IDA conducts systematic reviews of borrowers' public investment and expenditure programs and helps in determining investment priorities and rationalizing external aid. IDA also chairs many "consultative groups," which bring donors and borrowers together. The group meetings are often the dominant forum for discussions of the economic policies, development needs, and extent of external aid by the donor community.

Overview 4.7
The International Development Association

- IDA was established in 1960 as an affiliate of the World Bank to provide assistance to the world's poorest countries.
- IDA provides the poorest countries with interest-free loans—called credits—to finance projects that reduce poverty.
- IDA's financial resources come from contributions by donor governments including many wealthy and middle-income countries.
- IDA's annual lending is $5 to $7 billion.
- IDA's loan processing procedures are similar to those of the World Bank.
- IDA plays an important role in mobilizing and coordinating aid from other donors.

THE INTERNATIONAL FINANCE CORPORATION

The IFC was established in 1956 as an affiliate of the World Bank for the purpose of promoting private enterprise in the developing world. Unlike the World Bank, which lends money only to government entities, the IFC lends to private companies and may not accept guarantees of debt repayment from host-country governments. It also makes equity investments in private businesses. Quite often, project sponsors or their financial advisors or arrangers seek IFC participation not only because of the equity and financing it puts into a project but also because the IFC can mobilize additional loan and equity financing in the international markets.

The IFC brings other investors and lenders into a project in a variety of ways. It actively seeks partners for joint ventures and encourages other lenders to participate in projects it finances. Through its syndicated loans, the IFC is able to attract large amounts of commercial bank lending to companies in developing countries. The IFC also underwrites securities issues by companies in developing countries and helps in launching country funds.

Overview 4.8 The International Finance Corporation

- The IFC was established in 1956 to promote private enterprise in developing countries.
- Unlike the World Bank, the IFC lends directly to private companies, takes equity in private ventures, may not accept a government guarantee of debt repayment, and does not have strict procurement guidelines.
- Through its participation in a project, the IFC mobilizes other sources of investment and financing. In the first half of the 1990s, for every dollar invested by the IFC, six were provided by others.
- The IFC assists in establishing investment funds, tapping international bond markets, and developing local capital markets.

The IFC's direct contribution to financing projects in developing countries is significantly enhanced through its syndication of commercial bank loans. As a multilateral institution, IFC has certain privileges: it is exempt from payments of local taxes; its loans have never been rescheduled for political risk; it has access to high-level policymakers; and so on. When it syndicates a loan, the IFC is the lender of record and brings commercial banks under its own "umbrella." The commercial banks' loans thus are treated in the same way as the IFC's loans. Indeed, the IFC takes responsibility for administering the loans and collecting payments from the borrower. The IFC distributes all payments (whether received directly from the borrower or from realization of security) pro rata among the participants and itself. Thus, a default on any portion of the loan is a default to the IFC. The IFC also takes responsibility for appraising the project and coordinating the preparation of legal and contractual packages. With the IFC taking all the lead responsibilities, commercial banks find it more convenient and comforting to participate in financing projects in developing countries.

In the area of infrastructure, IFC's participation in project finance has been very effective in mobilizing other resources. During the first half of the 1990s, for every dollar invested by IFC a further six were provided by others. This is particularly important in the power sector, where the IFC had very little involvement before 1990 but has become an important player in the 1990s because of opportunities created by privatization and regulatory reform, particularly in Latin America (Argentina and Chile) and Asia (Pakistan, Philippines, and India).

The IFC's involvement in the oil and gas sector started in the late 1970s and grew at a moderate pace during the 1980s but surged markedly during the first half of 1990s, as an increasing number of countries facilitated private investments in the sector. The IFC invests in all parts of the oil and gas sector, upstream and downstream, except pure exploration. Where appropriate, it is prepared to participate in and lend to unincorporated joint ventures.

The IFC has been quite innovative in developing instruments for mobilizing debt and equity financing. Several of these instruments are of direct relevance to financing oil, gas, and power projects:

- The IFC's participation in build-operate-transfer (BOT) and build-own-operate (BOO) schemes has proved very useful in bringing in other investors and financiers and in resolving contractual, legal, technical, and financial issues with governments and government entities. Examples include power projects in Pakistan and the Philippines, gas and power projects in Argentina, power projects in Guatemala, and a power project in Chile.
- The IFC has taken an active role in the creation of specialized investment funds to mobilize resources from the world's major financial centers for lending and equity investment in developing-country projects.

An example is the Scudder Latin American Trust for Independent Power, formed in 1993. Four lead investors have committed a total of $100 million to this fund, which will invest in equity and target independent power producers in Latin America and the Caribbean. Another example is the Global Power Fund. The IFC has provided equity of $50 million in this fund, and two other founding shareholders have invested $200 million. The eventual target capitalization is $2.5 billion, to be raised from strategic investors, institutional investors, and possibly public equity issue. The fund would provide seed financing, including equity, subordinated debt, and completion guarantees, and bridge financing for power projects.

- The IFC assists private companies of developing countries in using international bond markets and international equity placements. The IFC helps companies design these instruments and provides loans and underwriting commitments. Examples are the underwriting of equity placement for an Indian power company, and co-management of an international bond issue for the Mexico City-Toluca Toll Road.
- The IFC has become involved in developing local capital markets. Again, the IFC provides broad technical assistance in this area. However, it also helps in mobilizing funds for specific projects. Examples are three private power projects in India with an estimated cost of $1.5 billion—a third of which is to be raised from domestic financial institutions.

Since the IFC cannot accept government guarantees of debt repayment, the success of its investments usually depends on the commercial success of the projects it helps finance. As a result, the IFC's project-processing procedure is more directly concerned with the proposed project than with the economy, the sector, and other broad policy matters, which are of concern to development institutions such as the World Bank and IDA. Nevertheless, the IFC's objective is not only to support private projects but also to encourage private investment and financing in the sector. It is therefore interested in improving the overall business environment and is careful not to compete with private sources of funds. Thus, an IFC appraisal pays attention to regulatory matters and availability of other sources of funds. In addition, the IFC follows the same standards of environmental safety as those of the World Bank. However, the IFC does not require borrowers to follow strict procurement guidelines but relies on good commercial practice.

In approaching the IFC for support, project sponsors should be aware that

- The IFC's interest rates and fees are determined on a commercial basis. The overall terms run from 7 to 12 years.
- The IFC will not finance more than 25 percent of the project cost for a greenfield project and can go slightly higher for project expansions.
- The IFC will take only a minority share in any entity.
- The IFC's shareholdings are usually treated as domestic or neutral capital for nationality ownership purposes.
- The IFC is a long-term equity investor but, when appropriate, it seeks to divest by selling its equity to local private investors either directly or through local stock exchanges.
- The IFC does not take an active role in the day-to-day management of a venture in which it participates.
- The IFC charges the project sponsors for the staff and resources it spends on preparing and appraising a project.

Overview 4.9
The IFC's Involvement in the Energy Sector

- The IFC had a presence in the oil and gas sector since the late 1970s.
- The IFC's involvement in private power projects has become substantial in the first half of the 1990s.
- The IFC's Global Power Fund channels private international funds into private power in developing countries.
- The IFC's appraisal processes are directly concerned with project risks, viability, and success.

THE MULTILATERAL INVESTMENT GUARANTEE AGENCY

MIGA was established in 1988 as an affiliate of the World Bank to encourage foreign investment in developing countries by providing investment guarantees against the risks of currency transfer, expropriation, war, civil disturbances, and breach of contract by the host government. MIGA's authorized capital is $1 billion, and its membership includes 152 developing and developed countries.

MIGA covers four categories of project risks, which can be purchased individually or in combination:

- Currency transfer coverage protects against losses arising from the investor's inability to convert local currency returns (profits, principal, interest, royalty, and so on) into foreign exchange for transfer outside host country. Coverage does not include currency devaluation.
- Expropriation coverage protects against partial or total loss of investment as a result of acts by the host government that may reduce or eliminate ownership or control of assets.
- War and civil disturbance coverage protects against losses from asset damage, destruction, or disappearance caused by politically motivated war or civil disturbance, including revolution, in the host country.
- Breach-of-contract coverage protects against losses arising from the host government's breach or repudiation of a contract with the investor.

MIGA's standard term is 15 years and covers both equity and loans. For equity, MIGA covers up to 90 percent of investment plus an additional 180 percent to cover earnings attributable to the investment. For loans, MIGA covers up to 90 percent of the loan and the interest that will accrue over the term of the loan. Rates vary according to the sector, project, and type of coverage.

MIGA has established itself rather well in the financing community during a short period. Its application-processing procedure is simple and efficient. However, its coverage is limited to $50 million per project and $150 million per country. Clearly, this is a small facility in the light of the financing needs of oil, gas, and power projects. Nevertheless, incorporating MIGA into a financing scheme facilitates mobilization of funds and guarantees from other agencies. Furthermore, MIGA may be able to provide greater coverage by syndicating insurance from other agencies.

THE ENERGY SECTOR MANAGEMENT ASSISTANCE PROGRAMME

After the oil crises of the 1970s, the United Nations Development Programme (UNDP) and the World Bank decided to form a joint program for assisting developing countries in dealing with energy crises. Initially called the Energy Assessment Programme, the effort concentrated on broad sector studies of energy development issues and strategies. After completing such studies for most developing countries, the Energy Assessment Programme began emphasizing follow-up studies on specific issues such as energy investment programs and energy pricing, along with subsectoral investigations in the areas of oil, gas, and power. In keeping with this shift, the name of the program was changed to the Energy Sector Management Assistance Programme (ESMAP).

ESMAP is a joint UNDP/World Bank effort, but it also has the financial support of 10 other donors, mainly European bilateral aid agencies. The donors provide about $10 million per year in the form of core and non-core funds. The core funds are paid by some donors in predetermined amounts each year. ESMAP management decides how to allocate these funds to various studies. The noncore funds are paid for specific studies. Thus, a proposal is prepared for each study and submitted to a donor for funding. Should the donor approve the proposal, the corresponding fund will be transferred to ESMAP to be allocated and spent on the proposed study. This is a time-consuming process.

ESMAP is managed by the World Bank on behalf of UNDP and the contributing bilateral donors. Although its budget is small, it is influential. ESMAP does not deal with detailed engineering or feasibility studies. Instead, it concentrates on resolving strategic and policy issues. These activities require a small budget but have a significant impact on governments' energy policies. During recent years, ESMAP has assisted a number of developing countries in privatizing their oil, gas, and power entities. It has also helped in setting up the appropriate regulatory framework for attracting the private sector in upstream oil and gas activities, independent power generation, and downstream petroleum and gas distribution. Since each ESMAP activity is, in practice, managed by World Bank staff, the results of the activity receive a high-level decisionmaking audience. In addition, because of World Bank relationships with government authorities, ESMAP study teams have easier access to the background policy matters than do consulting and other research groups.

Although ESMAP does not provide direct financing for project preparation, it can offer much to those who are trying to structure a project-finance package. The results of its past and ongoing activities are often relevant to various aspects of project preparation. For example, a major component of project preparation deals with projection of future regulatory changes in a country; possible adjustments to domestic petroleum, gas, and power prices; or expansions of gas and power grids. ESMAP activities deal with these issues based on first-hand information in each country.

Most ESMAP reports are publicly available. ESMAP staff can help project developers acquire in-depth information about strategic and policy issues in the oil, gas, and power sectors.

Overview 4.10 The Energy Sector Management Assistance Programme

- ESMAP is jointly sponsored by United Nations Development Programme (UNDP) and the World Bank. It also receives financial support from 10 bilateral donors, most of which are European, and is managed by the World Bank.
- Because of ESMAP's concentration on resolving strategic and policy issues and because of its access, through the World Bank, to high-level government authorities, the program has become an influential instrument in shaping countries' investment and sector strategies.
- In recent years, ESMAP has assisted particularly in restructuring and privatizing oil, gas, and power entities.
- ESMAP's past and ongoing studies contain useful information for the preparation of energy projects, especially on strategy and policy issues.

THE GLOBAL ENVIRONMENT FACILITY

The GEF is an important source of project financing. Its attraction is that it provides grant funds and concessional financing; its disadvantage is that it supports energy projects only under special circumstances.

As a financial mechanism, the GEF provides funds to developing countries for projects and programs that protect the global environment. Launched in 1991 as a pilot program, it was approved for continuous operation in 1994 with a core fund of about $2 billion.

The GEF is formally an international program with 90 member countries. Any member state of United Nations can become a member of the GEF. However, its financial resources are contributed mainly by the industrialized countries.

The resources of the GEF are available for projects and activities that address climate change, loss of biological diversity, pollution of international waters, and depletion of the ozone layer. In practice, the major portion of GEF resources are allocated to projects relating to global warming. Energy projects represent a significant component of the program.

The issue of global warming focuses on emissions of CO_2 and methane. Thus, projects that reduce these emissions are viewed as favorable to the global environment. For example, a decision to shift from coal-based to gas-based power generation would result in about 60 percent reduction in CO_2 emission. Likewise, a project aimed at rehabilitating gas pipeline systems would prevent gas leaks and reduce methane emissions.

The GEF encourages decisions that result in reduction of CO_2 and methane emissions. However, it provides financial support only to projects that are *not* economically viable based on their own costs and benefits but *become* economic when global environmental benefits are considered. For example, there is substantial international sympathy for preventing flaring of natural gas for a variety of reasons, including benefits to the global environment. Nevertheless, in some countries the cost of collecting the gas exceeds the benefit of using it. Such a project would not be economically viable if assessed on the basis of costs and benefits to the project sponsors (or the country involved). However, if the favorable impact on the global environment was assessed and added to the benefits, then the project might become viable. The problem that would normally arise is that the benefit to the global environment would not be received by the project sponsors or the country, which therefore would not have an incentive to undertake the project. This is where the GEF can assist—as it did recently in a project to reduce gas flaring in India—by compensating project sponsors (or the country) with incremental funding to make the project viable. In practice, an assessment is made of the difference between the costs of a project undertaken with global environmental objectives in mind and the costs of an alternative project that the country would have implemented in the absence of global environmental concerns. The GEF pays this difference.

The operation of the GEF is shared among the UNDP, the United Nations Environment Programme (UNEP), and the World Bank. The UNDP is responsible for technical assistance. The UNEP is responsible for catalyzing development of scientific and technical analysis. The World Bank, the repository of the GEF Trust Fund, is responsible for selecting and monitoring investment projects. Thus, in order to receive support for a project, sponsors need to work with the GEF at the World Bank.

Overview 4.11 The Global Environment Facility

- The Global Environment Facility (GEF) is a financial mechanism that provides grant and concessional funding for projects that have a favorable impact on the global environment. In the area of global warming, the main concerns are emissions of CO_2 and methane. Energy projects such as pipeline rehabilitation, collection of flared gas, and other efficiency improvement investments that result in a reduction in these emissions are viewed as favorable to the global environment.
- A project would be eligible for GEF financial support only if it is not viable based on its costs and benefits to the country but becomes viable when global environmental benefits are incorporated into the cost-benefit analysis.

5

Borrowing from Regional Development Banks

Most regional development banks were created in the 1950s and 1960s with objectives similar to those of the World Bank, but focused on specific regions. Also like the World Bank, the regional banks were designed to assist in reducing poverty and in promoting economic growth. Moreover, they were modeled after the World Bank with respect to procedures for project preparation and implementation.

The regional development banks are owned by governments of the corresponding regions and by governments of industrialized countries.

THE INTER-AMERICAN DEVELOPMENT BANK

The Inter-American Development Bank (IDB) is the oldest of the regional development institutions. It was established in 1959 to help economic and social development in Latin America and the Caribbean. The IDB's membership, originally limited to Latin American countries and the United States, now includes 17 nonregional, mainly industrialized, countries in its total of 46 nations.

IDB's financial resources consist of ordinary capital resources (subscribed capital, reserves, and funds raised through borrowing) and contributions made by member countries. Member countries' subscriptions to the capital fund consist of paid-in and callable capital. The paid-in subscription is a cash payment and represents only 2.5 percent of each member's subscription. Thus, the majority of the capital funds is callable capital; that is, it comprises guarantees by member governments to provide funds if the IDB cannot repay what it borrows on the capital markets of Europe, Japan, Latin America, and the United States.

The IDB's annual lending is about $5 to $7 billion and covers all social and economic sectors. It lends to public entities and requires a government guarantee. A major part of the total lending (20 to 25 percent) is allocated to the energy sector. The IDB's lending strategy initially emphasized financing social projects and helping the poor. In the 1990s, however, some of the emphasis shifted to private sector development and assistance for environmental protection and cleanup. The emphasis on private sector development increased the number of sector loans. These loans, which are policy-driven and disbursed rapidly, are intended to promote restructuring and reform and to create the right business environment for private sector investments.

Overview 5.1 The Inter-American Development Bank

- The Inter-American Development Bank (IDB) is the oldest regional development institution. Founded in 1959, it has 46 member countries, of which 17 are nonregional industrialized countries.
- The IDB's annual lending is $5 to $7 billion, of which 20 to 25 percent goes to the energy sector. The IDB lends to public entities and requires a government guarantee. It also has a grant program for supporting project preparation, sectoral studies, and other technical assistance activities.
- In order to provide direct support to the private sector, two IDB affiliates—the Inter-American Investment Corporation (IIC) and the Multilateral Investment Fund (MIF)—have been formed.
- The IIC provides equity and loans to private companies without a government guarantee. IIC participation is limited to $6 million per project, but it can arrange syndicates to mobilize commercial bank loans.
- The MIF supports both foreign and domestic investment by advancing market-oriented policy initiatives and reforms, financing technical assistance in investment-related areas, and supporting training and institution building.

To provide more direct support for private sector investment, the Inter-American Investment Corporation (IIC) was formed as the private sector affiliate of the IDB. IIC operations may take the form of loans, investment in stock capital, or guarantees. Eligible projects may include new companies or the expansion, modernization, restructuring, or privatization of existing ones. The IIC finances up to 33 percent of project cost. It will hold no more than one-third of the share capital. It will not assume administrative or managerial duties in the companies. The IIC's total direct participation in a project is limited to $6 million. This small ceiling, imposed by the IIC's capitalization status, limits IIC involvement in most energy projects. However, because of its multilateral status, the IIC's presence in a project facilitates equity and debt financing by other sources. The IIC itself takes responsibility for arranging syndicated loans. Under such arrangements, IIC provides a large loan, which in effect is made up of contributions from syndicating commercial banks. The IIC remains lender of record, serves as administrator of the loan, and conducts the project's feasibility studies. A default on any portion of the loan would be treated as a default to IIC. In addition to the greater security for commercial bank participants, IIC syndication arrangements provide additional advantages, such as tax exemptions and a closer working relationship with host governments. The IIC's projects in the energy sector are primarily in private power generation.

THE ASIAN DEVELOPMENT BANK

The Asian Development Bank (ADB) was formed in 1966. Its membership includes 52 countries (36 from Asia and 16 from outside the region, of which most are industrialized). The ADB's ordinary capital consists of subscribed capital, reserves, and borrowed funds. This subscribed capital consists of paid-in capital, paid-in cash by member countries, and callable capital (guarantees by member countries to provide funds in the event that ADB cannot repay funds it has borrowed on the capital markets). The paid-in capital is 12 percent of total subscribed capital. The borrowed funds are raised in Europe, Asia, the Middle East, and the United States.

In addition to ordinary capital, the ADB has funds available through a special fund, the Asia Development Fund, which receives contributions from rich member countries. This fund lends on concessionary terms to very poor Asian nations. Loans from this fund are interest-free and have maturities of up to 40 years. The ADB's total annual lending is about $5 billion, of which 25 to 30 percent goes to the energy sector. Lending has primarily been to traditional public projects, particularly power generation, transmission, and distribution projects. In its traditional lending, the

ADB requires a government guarantee. However, since the 1980s it has emphasized private sector participation. Support for the private sector increased in 1983 with the introduction of an equity investment facility, which allowed the ADB to make direct equity investments in private enterprises and financial institutions and to extend lines of equity to selected financial intermediaries. In 1985, ADB further reinforced its support for the private sector by establishing a facility for direct lending to private enterprises without government guarantee.

The ADB has been very supportive of build-operate-transfer and build-own-operate projects in the power sector. For these projects, it provides equity and loans with and without government guarantee. This is particularly useful for projects in which a public entity holds a minority share. The ADB is then able to finance the public entity's equity and debt share through a loan with government guarantee and the private sector's equity and debt share through direct participation.

Overview 5.2 The Asian Development Bank

- The Asian Development Bank (ADB) was founded in 1966. Its members include 36 Asian countries and 16 industrialized countries from outside the region.
- The ADB's annual lending is about $5 billion, of which 25 to 30 percent goes to the energy sector. It traditionally lends to public entities that have a government guarantee.
- The Asia Development Fund is a special fund of the ADB that lends on concessional terms to poor Asian nations.
- In the 1980s, ADB created an equity investment facility and a private sector lending facility. Through these facilities, the ADB provides equity and debt financing without government guarantees.
- The ADB provides grant funds for technical assistance that can be utilized for certain aspects of project preparation.

The ADB also provides grants that can be used to study strategic issues and policies related to sector management and investments. Many of these grants have been used to carry out studies in the energy sector, which have ranged from broad sector issues to specific preparatory work for significant investments in gas and power. The ADB also has at its disposal a substantial grant facility funded by the government of Japan and other facilities furnished by the governments of other Organisation for Economic Co-operation and Development (OECD) countries.

THE AFRICAN DEVELOPMENT BANK

The African Development Bank (AfDB) was formed in 1963. Its membership includes 51 African states and 25 other nations, most of which are industrialized. The AfDB Group refers to the AfDB itself, the African Development Fund, which was established in 1972, and the Nigerian Trust Fund, which was established in 1975. The AfDB is the most significant entity in the group and provides about two-thirds of its total lending. It borrows on capital markets and lends to the more developed nations of Africa. Project loans have maturities of up to 20 years at interest rates that are tied to the cost at which AfDB borrows its funds.

Overview 5.3 The African Development Bank

- The African Development Bank (AfDB) was formed in 1963. Its membership includes 51 African states and 25 other nations.
- The AfDB provides loans at rates tied to its own cost of borrowing on capital markets.
- An AfDB affiliate, the African Development Fund, receives financial contributions from nonregional member countries and provides concessional loans to the poorest African countries.

The African Development Fund is a concessional fund similar in concept to International Development Association (IDA, of the World Bank Group). Contributors to the African Development Fund are the AfDB's nonregional members. The fund, managed by AfDB, provides low-cost loans to Africa's poorest nations. Loan maturities are up to 50 years, and no interest is levied except for a 0.75 percent service charge.

The Nigerian Trust Fund, which is fully funded by the Nigerian government, is another part of the AfDB Group, although its capitalization and range of activities are comparatively minor.

The AfDB Group's annual lending is $3 to $4 billion, of which about 20 percent is invested in the energy sector. Like other multilaterals, AfDB supports private sector investments. In 1991 it established a Private Sector Development Unit (PSDU), which operates as a separate organizational entity. The PSDU mobilizes financial resources for private sector projects through term loans and equity participation. PSDU loans and investments do not require government guarantee.

THE EUROPEAN UNION

In the Treaty of Rome of 1957, the six signatory countries created the European Economic Community (EEC). After merging with the European Coal and Steel Community and the European Atomic Energy Community in 1967, it became known as the European Communities or the European Community (EC). The European Union (EU) was created in 1993 with the entry into force of the Maastricht Treaty on European Union, which incorporated the EC and two new formalized areas of intergovernmental cooperationæsecurity and foreign policy and justice and home affairs. The aim of the EU is to integrate member economies fully. The main institutions of the EU are the Council of Ministers, the European Parliament, the Court of Justice, and the European Commission. The EU exercises supranational powers and maintains its own unit of account, the European Community Unit (ECU). The membership has now expanded to include 15 of the most industrialized European countries.

The EU pays special attention to its relationship with developing countries. It has signed a large number of international agreements with developing countries involving the full range of aid and trade development instruments and maintains a permanent dialogue with them on policy issues and cooperation.

EU assistance has been extended to countries in Africa, the Caribbean, the Pacific, the southern and eastern Mediterranean areas, Latin America, Asia, central and eastern Europe, and the former Soviet Union (FSU).

EU assistance to developing countries is $5 to $6 billion per year, all of which is in the form of grants. Sectors of interest include energy. The EU also contributes to various international programs, such as the Special Program of Assistance for Africa and most of the assistance programs to eastern Europe and the FSU.

THE EUROPEAN INVESTMENT BANK

As part of the decision to establish the EEC, the European Investment Bank (EIB) was set up in 1958. The objective was to finance capital investment projects that promote balanced development throughout the community. The EIB is owned by the 15 member countries of the EU. Its subscribed capital consists of 7.5 percent paid-in capital; the rest is callable capital. It operates as a bank, raising the bulk of its financial resources on capital markets. It lends the proceeds of its borrowings to finance projects on a nonprofit basis at cost plus 0.15 percent to cover administrative expenses.

The volume of the EIB's operation is about $15 billion a year. Although the bulk of its loans are within the EU, the EIB has been called on to participate in the community's development cooperation activities. Within this policy framework, the EIB operates outside the community in 12 Mediterranean and 69 African, Caribbean, and Pacific states. It has also financed projects in Poland and Hungary. Lending to countries outside the community accounts for about 10 percent of total lending. EIB loans are linked to public and private investments in infrastructure, industry, agro-industry, agriculture, energy, tourism, and service sector projects. The terms of EIB loans are up to 10 to 12 years for industrial projects and 12 to 15 years for infrastructure and energy projects.

THE EUROPEAN BANK FOR RECONSTRUCTION AND DEVELOPMENT

The European Bank for Reconstruction and Development (EBRD) was established in 1991 to foster the transition of central and eastern European economies toward open, market-oriented operation and to promote private initiative in central and eastern European countries. Membership includes 52 countries plus the EU and the EIB. The EU member countries and EIB subscribe 51 percent of the authorized capital, which is 10 billion ECU, the EBRD's unit of account (1 ECU = 1.2 US$). Among other members, the United States subscribes 10 percent, Japan 8.5 percent, and Canada 3.4 percent. The EBRD's subscribed capital consists of 30 percent paid-in and 70 percent callable capital. It also borrows on international capital markets to meet member countries' demand for debt and equity funds.

The EBRD offers both merchant banking and development banking services by providing loans, equity investments, and technical cooperation. Furthermore, it often acts as a catalyst for attracting additional resources.

Overview 5.4 The European Investment Bank and The European Bank for Reconstruction and Development

The European Investment Bank (EIB) and European Bank for Reconstruction and Development (EBRD), though somewhat related, were formed with entirely different objectives:

- The EIB was set up in 1958 to finance investments in the six member countries of the European Economic Community (EEC), which became the European Community (EC) in 1967 and then the European Union (EU) in 1993 (it now has 15 members). It has on occasion provided support to projects in developing countries. From an annual lending of about $15 billion, about 5 percent normally flows to developing countries.
- The EBRD was established in 1991 to assist the transition of central and eastern European countries to market-oriented economies. Its membership includes 52 countries plus the EU and the EIB. The EBRD provides support in the form of equity participation and debt financing. Its lending rate is set at a spread over LIBOR.

Merchant banking operations (about 60 percent of EBRD operations) favor projects that typically have a foreign industrial partner and a few foreign and local banks participating in the financing plan. EBRD also supports projects with no foreign partner. In addition, it has established or participated in capital investment funds and has invested in local banking systems.

EBRD's development banking operations (about 40 percent of total operations) address the lack of institutional and legal framework in borrowing countries as well as the deficiencies in public and financial infrastructure.

The EBRD lends at spreads over a market benchmark, usually the London Interbank Offer Rate (LIBOR). It also seeks to provide borrowers with other capital-market products, such as caps and swaps (as long as EBRD itself is fully hedged and thus passes on the benefit of its rating). The EBRD offers fixed-rate and floating-rate loans in dollars, deutschemarks, and ECUs, as well as multi-currency loans. The EBRD has helped recipients raise their own capital in a number of cases, and provided nonrecourse debt, equity, and guarantees. Maturities are from 5 to 10 years for operations in the private sector and can be extended up to 15 years for public infrastructure projects. Procurement of goods and services is open worldwide.

THE NORDIC INVESTMENT BANK

The Nordic Investment Bank (NIB) was established by the five Nordic countries of Denmark, Finland, Iceland, Norway, and Sweden in 1975 as a multilateral financial institution to finance investments of interest to its members.

NIB provides Nordic Loans for projects within the Nordic countries and Project Investment Loans (PILs), on an international basis. PILs are provided where projects pass a "Nordic Interest Requirement"; that is, they have a positive effect on the Nordic region—for example, by involving deliveries of goods and services from at least two Nordic countries.

Assistance to countries outside the Nordic countries accounts for about one-third of NIB's $500 million per year lending total. Sectors of interest include energy. NIB loans are extended on commercial terms.

THE NORDIC DEVELOPMENT FUND

The Nordic Development Fund (NDF) became operational in February 1989 as a multilateral financial institution funded by the aid budgets of the Nordic countries. The NDF provides long-term credits to low- and lower-middle income developing countries on concessional terms to promote social and economic development. The NDF engages only in cofinancing and builds its portfolio on operations initiated by other institutions. It has a limited staff. The NDF's cofinancing partners are the World Bank, IFC, IDB, AfDB, ADB, and NIB. Sectors of interest include energy. Countries of interest are low-income developing nations, primarily in Africa and Asia.

The NDF's credits are granted on the same terms to all eligible countries. The terms are concessional—that is, interest-free—with an annual service charge of 0.75 percent. Credits have 40-year maturities, including a 10-year grace period. They are generally limited to $5 to $6 million per project.

NDF funds are primarily intended for procurement from Nordic sources, under Nordic competitive bidding procedures. When consultant services are financed, the NDF encourages cooperation between Nordic and local consultants and sometimes finances certain local services.

THE OPEC FUND FOR INTERNATIONAL DEVELOPMENT

The OPEC Fund for International Development is an autonomous multilateral development institution, established in 1976 by the 13 (now 12) members of the Organization of Petroleum Exporting Countries (OPEC). Its objectives are to promote cooperation among OPEC member countries and other developing countries by providing financial assistance in support of economic and social development.

Most of the fund's assistance is in the form of project loans; however, some balance-of-payments support and structural adjustment programs are offered. Outright grants are extended in support of technical assistance, food aid, and research. Emergency relief aid is provided occasionally.

Geographical interests include most countries in Africa and Asia and a few countries in Latin America. Sectors of interest include energy. The annual level of assistance is about $200 million, of which an average of 30 percent is allocated to energy projects. The assistance includes loans and grants. In the case of loans, terms are decided on a case-by-case basis depending on the project and the recipient country.

THE ISLAMIC DEVELOPMENT BANK

The IsDB is an autonomous multilateral development institution, established in 1974 by member countries of the Islamic Conference Organization. It includes 45 countries in Africa, Asia, and the Middle East.

The objectives of the IsDB are to contribute to the economic development and social progress of member countries and Muslim communities in non-member countries, in accordance with the principles of Islamic Sharia. It provides assistance through interest-free loans for development projects and enterprises and finances leasing, installment sales, equity investments, foreign trade, technical assistance, and research. There are also funds for special purposes, including trust funds and funds for Muslim communities in nonmember countries.

The countries of interest include very poor Muslim nations, primarily in Africa. Assistance is extended in a variety of forms. The following services are of some relevance to the energy sector:

- Loans, used to finance infrastructure projects. These are interest-free, carrying an annual service fee ranging between 1.5 and 2.5 percent. Maturities range from 15 to 25 years, including a grace period of 3 to 5 years.
- Technical assistance, provided in the form of grants or loans mainly for the identification and preparation of investment projects, including feasibility studies and institution building. Such assistance is primarily aimed at IsDB's least-developed member countries.
- Special programs, mainly in the form of grants for research, vocational, and administrative training; educational programs; disaster relief; and assistance to Muslim communities in nonmember countries.

Procurement is subject to international competitive bidding (ICB).

THE ARAB FUND FOR ECONOMIC AND SOCIAL DEVELOPMENT

The Arab Fund for Economic and Social Development is a regional development institution established in 1972 at the initiative of the Arab League. Located in Kuwait, it seeks to assist the 21 member countries of the Arab League in their efforts toward economic and social development. The fund's activities include financing development projects, participating in equity investments in private and public enterprises, and providing technical assistance grants.

The largest recipients of funds are Morocco, Egypt, Syria, and Tunisia. However, most small African countries receive assistance. The average level of lending is $200 to $300 million per year. Energy projects account for 20 to 30 percent of the total. The average lending terms include an interest rate of 4 percent and maturities of 23 years, with grace periods of 5 years. Procurement is subject to international competitive bidding.

THE ARAB BANK FOR ECONOMIC DEVELOPMENT IN AFRICA

BADEA (the French acronym for the Arab Bank for Economic Development in Africa) is an autonomous regional development institution. BADEA provides financial and technical assistance to African countries for economic development and promotes the movement of Arab capital toward African countries.

Established by the Arab League in December 1973, BADEA began operations in March 1975. Its member countries are Algeria, Bahrain, Egypt, Iraq, Jordan, Kuwait, Lebanon, Libya, Mauritania, Morocco, Oman, Palestine, Qatar, Saudi Arabia, Sudan, Syria, Tunisia, and United Arab Emirates. Any Arab country is eligible to join BADEA.

All member countries of the Organization of African Unity (OAU) that do not belong to the Arab League are eligible to apply for assistance. Almost every eligible African country has benefited from BADEA assistance.

Sectors of interest include energy. Lending terms have tightened progressively over the years. The overall grant element of commitments has fallen progressively. The rates of interest range from 2 to 7 percent, maturities from 10 to 25 years, and grace periods from 3 to 6 years, depending on the kind of project and the economic situation of the borrower. Procurement is subject to international bidding procedures. A preference of up to 10 percent is given to bidding enterprises fully or partially owned by Arab or African interests.

6

BILATERAL SOURCES OF FINANCING

Bilateral agencies are development institutions set up in industrialized countries to support the investment and technical assistance requirements of developing countries. These agencies normally receive part or all of their funds from their respective governments. Their support for developing countries is quite often linked to the promotion of sales of goods and services supplied by their own countries' companies.

The functions of bilateral agencies fall into two distinct categories:

- Provision of grants and highly concessional loans to developing countries based on economic, social, and political considerations.
- Provision of loans, guarantees, and insurance that are designed to help exportation of goods and services from the donor country and to promote the involvement of donor-country companies in projects in developing countries.

In most donor countries, separate institutions have been set up to handle the two functions. For example, in the United States, the U.S. Agency for International Development (USAID) is primarily responsible for development assistance, and the U.S. Export-Import Bank (USExim) promotes U.S. business abroad. However, in practice the functions of the two agencies are intertwined. Furthermore, other public and private entities become involved, which results in overlap of activities within each donor country.

DEVELOPMENT ASSISTANCE

All Organisation for Economic Co-operation and Development (OECD) countries have substantial development assistance programs. In volume, Japan and the United States represent the largest donors of development funds. However, in relative generosity, European countries make the larger contributions. For example, Norway, Sweden, Denmark, and the Netherlands contribute about 1 percent of their GNP, whereas Japan contributes about 0.3 percent of its GNP and the United States about 0.15 percent of its GNP. The figures for France, Germany, and the United Kingdom are 0.6 percent, 0.4 percent, and 0.3 percent, respectively.

The objectives of development assistance programs vary substantially. U.S. assistance has a strong focus on countries and regions the United States considers important to maintaining international stability. The Japanese program is aimed at strengthening economic relations. France's assistance is based on strong historical ties. Many other European countries base their development assistance on social considerations.

The themes of development assistance programs are somewhat different among donor countries, but most now have two themes in common. First, a priority for most programs is provision of appropriate institutional structures for private sector development. Second, most donors require preservation of the environment and incorporate strict standards for protection of the environment into their assistance programs.

Development assistance programs and agencies, as parts of governments, are funded through public budgets. The funds are provided to developing countries in the form of grants or highly concessional loans. Some of these funds are channeled to developing countries through multilateral institutions—for example, the World Bank, regional development banks, and United Nations agencies. The rest of the funds are provided to developing countries on a bilateral basis.

Bilateral assistance programs provide funding to developing countries in support of (1) balance-of-payment requirements (that is, the money will be at the disposal of the government of the recipient country to be disbursed against general imports); (2) specific investments (for example, building a power plant or a pipeline); (3) preparation of a project (for example, feasibility studies); and (4) more general technical assistance, such as strategic studies and training.

The last two items have become important components of bilateral aid programs. Most donor countries have established separate entities to handle the technical assistance program or made it a major responsibility of a related entity.

EXPORT FINANCING

All OECD countries and some developing countries—such as Korea, India, and Brazil—provide concessional financing for export of goods and services by their own companies. The financing is provided either as a loan to the purchaser of the equipment (for example, a power company in a developing country) or as a credit to the supplier (for example, a U.S. manufacturer of power equipment). The loan can be made directly by the bilateral agency to either party or through a financial intermediary, such as a commercial bank. It can be made by a commercial bank while the bilateral agency insures repayment by the borrower.

Provision of export credit represents an important dimension in competition among the equipment suppliers of various countries. The presence and extent of concessional finance sometimes distort price competition and may discourage efficiency of manufacturers. In order to avoid undesirable distortions, the governments of OECD countries agreed in 1978 to the "Arrangement on Guidelines for Officially Supported Credits," known less formally as the OECD Consensus. The arrangement sets limits on the favorability of terms of official export credit. All OECD countries agreed not to provide export credit exceeding 85 percent of the contract value and not to set interest rates below the OECD interest rate matrix, which is revised semi-annually.

In the energy sector, agencies such as USExim and the Export-Import Bank of Japan (JExim) play important roles in financing power and gas projects. Loans are normally made to the purchaser of equipment and services. Three methods are used to extend finance to an importing entity:

- Direct lending, where the bilateral agency signs a loan agreement with the importing entity. These loans are normally tied to the purchase of equipment from the bilateral agency country's manufacturers. The terms follow the OECD Consensus. JExim also provides untied loans, where the borrower can purchase equipment and services through international competitive bidding. Terms of untied loans do not have to follow the OECD Consensus. The direct loans (tied or untied) are normally somewhat political and are decided by high-level government authorities of exporting and importing countries.
- Lending through financial intermediaries (also referred to as *relending* or *onlending*), where the bilateral agency provides the loan to a commercial bank, which relends to the importer.

- Interest rate equalization, where a commercial bank provides a loan to the importer at a lower-than-commercial rate (normally the OECD Consensus rate). The commercial bank is then compensated by the corresponding bilateral agency for the difference between the commercial and consensus rates.

The idea behind the second and third methods is to involve commercial banks in export financing. Many countries, but most notably Japan, provide direct loans through negotiations between corresponding governments and within the frameworks of their overall aid programs to recipient countries.

Overview 6.1 Objectives of Bilateral Aid Agencies

- All Organisation for Economic Co-operation and Development (OECD) countries have set up agencies that provide financial and technical assistance to developing countries. These agencies provide grants and concessional loans to support general import requirements (balance of payment) or specific programs and projects.
- Their support is aimed at assisting developing countries and sometimes aimed at promoting exports and other business interests for their own nationals.
- Most OECD countries have separate agencies for providing development assistance and export promotion. However, functions of these two agencies quite often overlap, and their funds may be combined to provide attractive financing packages.
- Export credit agencies often provide guarantees for loan repayment and insurance against political risks and other risks to commercial lenders. Some countries have separate agencies responsible for risk insurance.

INSURANCE AND GUARANTEES

In addition to providing support for developing countries, insurance and guarantees from bilateral agencies are aimed at promoting the sale of goods and services by the lending country's companies. Export credit insurance protects the exporter or bank against commercial or political risks.

Commercial risks include nonpayment by a nonsovereign or private sector buyer in the event of insolvency or failure to take shipments of goods. Protection is also available against preshipment risks and unfair calling of the exporter's commitment bonds.

Political risks include default by a sovereign entity, general moratorium on external debt, foreign exchange inconvertibility, cancellation of export-import licenses, delay in transfer of payments, war, and civil disturbances.

Figure 6.1 Export Credit, Insurance and Guarantee Facilities of Bilateral Agencies

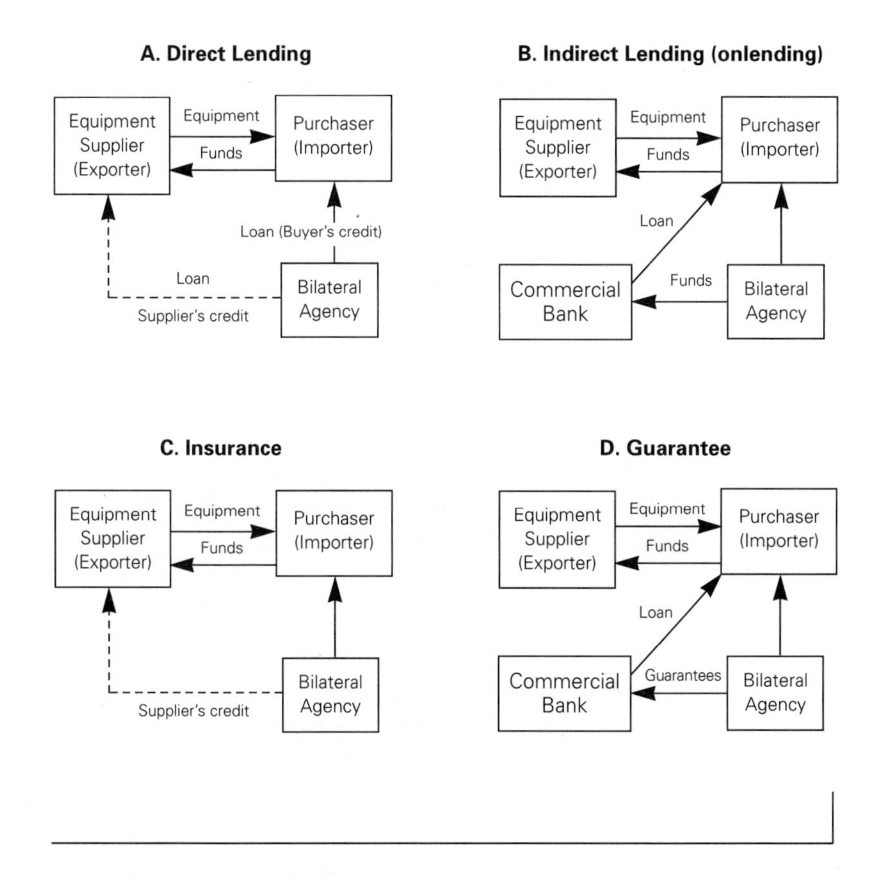

Insurance and guarantees from bilateral agencies normally cover 85 to 95 percent of losses caused by commercial risks and 100 percent of losses caused by political events. Coverage is either global or specific. Global coverage provides protection for all export transactions of a supplier or bank. Specific coverage provides protection for one transaction. Premiums for global coverage are lower than for specific coverage because of the spread of risks.

Overview 6.2 The Major Bilateral Development Assistance Agencies

- The largest contributors to bilateral development assistance are Japan, the United States, France, and Germany. Among other OECD countries, Canada, the United Kingdom, Italy, Finland, Norway, Netherlands, Sweden, Switzerland, Belgium, Denmark, Australia, and Austria make significant contributions.
- A few non-OECD countries provide development assistance, such as Kuwait, Saudi Arabia, Abu Dhabi, and Korea, and have export credit and support facilities, such as India and Brazil.

Japan

Japan is the largest provider of development assistance. The program of Japanese assistance to developing countries provides financial and technical assistance to less developed nations and recycles Japan's surplus funds to these countries. The bilateral aid program includes an official development assistance component and an "other official flows" category, which refers to loans under the Capital Recycling Plan.

Under a 1993 initiative, the Japanese government allocated $120 billion of untied funds for assistance to developing countries over the five-year period from 1993 to 1998. This translates into an average of $24 billion per year, almost twice the annual U.S. development assistance program.

Several government ministries are involved in planning, budgeting, and allocating development assistance funds. The Ministry of Foreign Affairs

Table 6.2 Major Agencies Involved in Bilateral Financing

Country	Development Assistance Agency	Export Credit Agency	Insurance and Guarantee	Other Involved Parties
Japan	OECF, JExim, JICA	JExim	EID	JNOC, JCI
United States	USAID	USExim, PEFCO	USExim, OPIC	TDA, PSED
France	CFD, FAC	BFCE	COFACE	DREE
Germany	BMZ, KfW, GTZ	KfW, AKA	Hermes	DEG
U.K.	ODA, CDC, Crown Agents	ECGD, DTI	ECGD	
Italy	DGCS	MEDIOCREDITO	SACE	
Canada	CIDA	EDC	EDC	
Netherlands	DGIS		NCM	
Denmark	DANIDA	EFC	EKR	IFU
Belgium	BADC	Creditexport, Copromex	OND	
Finland	DIDC (FINNIDA)	FEC	FGB (VTL)	FINFUND
Austria	DGDC	OEKB	OEKB	
Norway	NORAD	Exportfinans	GEIK	
Sweden	SIDA	SIDA	EKN	
Switzerland	DEH	BAWI		
Spain	MOEF	ICEX	CESCE	AIC
Portugal	FCE			IEC
Australia	AusAID	EFIC	EFIC	
Kuwait	Kuwait Fund			
Saudi Arabia	Saudi Fund			
Abu Dhabi	Abu Dhabi Fund			
Korea	KExim, EDCF	KExim		KOICA

(MOFA) receives all requests for bilateral grants and loans. MOFA is responsible for extending and administering grants. The Ministry of Finance (MOF) is in charge of the budget and works in consultation with other offices to determine allocation of loans and grants. MOF is also involved in the allocation of grants for technical assistance.

The main agencies responsible for implementation of the development assistance program are

- The Overseas Economic Cooperation Fund (OECF), responsible for identification, appraisal, negotiation, and implementation of concessional loans in the category of official development aid.
- JExim, responsible for the processing and implementation of the "other official flows" loans.
- The Japan International Cooperation Agency (JICA), the oversight agency for technical grants and technical cooperation, including provision of equipment, training, technical assistance services, and development surveys.

About 80 percent of the OECF's loans are provided on a fully untied basis; that is, procurement is not tied to Japanese export. JExim also has an Untied Loan Facility, which provides direct loans to foreign governments and financial institutions for funding of development projects and programs. The OECF and JExim are both very active in the energy sector. The former provides soft loans for major infrastructure projects to countries that are unable to use commercial loans for such projects. The average annual interest rate is 2.6 percent. The maturity is 25 to 30 years, including a 7- to 10-year grace period. Terms depend on the recipient's per capita income, the country's political relationship with Japan, and project profitability. JExim provides untied loans with semicommercial interest rates. They can include a grant component of up to 25 percent. The maturity is 10 to 20 years, including a 3- to 5-year grace period.

For historical, geographic, and economic reasons Asian countries account for the majority of OECF and JExim untied loans. In regard to OECF, Asia accounts for about 80 percent of total commitments, followed by Africa, 8 percent; Latin America, 6 percent; and the Middle East, 5 percent. Regarding JExim untied loans, Asia accounts for 43 percent; followed by Latin America and the Caribbean, 22 percent; and eastern Europe, Africa, and the Middle East, 16 percent.

Export support is provided primarily through two agencies: JExim and the Export Insurance Division (EID) of the Ministry for International Trade and Industry (MITI). JExim facilities include (1) export credits to Japanese exporters as well as foreign buyers, banks, and governments and (2) direct loans to joint ventures involving Japanese corporations and to foreign governments (for capital contributions to joint ventures with Japanese corporations). The EID provides insurance against commercial and political risks related to export financing. It also provides insurance for investments by Japanese nationals in other countries.

Overview 6.3 Japan's Bilateral Assistance Program

- Japan is the largest provider of assistance to developing countries.
- The main aid agencies are the Overseas Economic Cooperation Fund (OECF), the Export-Import Bank of Japan (JExim), and the Japan International Cooperation Agency (JICA).
- Export support is provided through JExim and the Export Insurance Division (EID) of the Ministry of International Trade and Industry (MITI).
- Financial support for preinvestment studies, training, and other types of technical assistance is provided by JICA, OECF, and JExim.
- A source of support for oil exploration and production projects is the Japan National Oil Corporation (JNOC).

Financial support for technical assistance—for example, pre-investment studies and training—is provided by both the OECF and JExim. In addition, JICA is the official agency designated by the Japanese government to extend technical cooperation to developing countries. Its functions include funding sectoral and project-specific studies, providing expertise and equipment aimed at transfer of technology, and mobilizing finance from the OECF and JExim for underlying investments. Another agency operating in this area is the Japan Consulting Institute (JCI), which provides technical assistance for project preparation but does not provide funds for carrying out studies and preparatory work.

A source of support specifically for oil exploration and production projects is the Japan National Oil Corporation (JNOC). JNOC is an independent government body that provides finance either in the form of equity or direct loans to Japanese companies investing in oil exploration and development. Because oil potential in Japan is limited, JNOC primarily supports overseas projects. JNOC provides funds for exploration. If companies fail to make commercial discoveries they can be exempted from repayment of JNOC loans. If exploration is successful and companies need funds for development, JNOC guarantees repayment of loans made by commercial banks. JNOC often negotiates terms of exploration investments with governments of host countries on behalf of Japanese oil companies.

United States

The U.S. development assistance program, about $12 billion annually, primarily pursues political objectives. It is the second-largest bilateral program (after Japan). More than 60 percent of the assistance takes the form of balance-of-payments support.

The policymaking and implementation of the development assistance program involve the following agencies:

- USAID, the leading agency for bilateral assistance. It has offices in about 100 developing countries.
- USExim, which provides loans and loan guarantees, independently and in cooperation with other lenders, for procurement of U.S. goods and services.
- The Overseas Private Investment Corporation (OPIC), which provides project financing through loans and loan guarantees, as well as political risk insurance and advisory services in support of U.S. investment and exports. OPIC has a special Project Development Program for central and eastern Europe and the former Soviet Union (FSU).
- The U.S. Trade and Development Agency (TDA), which provides funding for large project feasibility studies by U.S. firms.
- The U.S. Peace Corps, which provides grassroots assistance in approximately 70 developing countries.
- The U.S. State Department, which coordinates foreign policy aspects of U.S. assistance.
- The U.S. Treasury Department, which handles the interactions with multilateral agencies.

In addition to providing grants for financing projects, USAID has numerous programs of support for investment studies and preparatory work. Many of these programs are aimed at providing assistance to the energy sector and have no direct link to promoting the interest of American companies; others support projects that do have potential business opportunities for U.S. companies. In the first category, studies supported by USAID have helped many developing countries to formulate energy development strategies, energy conservation and pricing policies, and so on. In the second category are several country-specific programs that provide funds for work promoting joint ventures between each specific country and American companies. Examples are programs with India, Egypt, and the Philippines. Also in this category are two important programs—the TDA and the Private Sector Energy Development Program (PSED).

The TDA provides grants for prefeasibility and feasibility studies for proposed projects in developing countries. The grant must be requested by a government entity in the developing country, but the work must be undertaken by a U.S. company. In reviewing grant requests, the TDA considers whether the study is likely to identify potential large-scale export opportunities for U.S. companies. Grants are normally between $500,000 and $1 million for studies in the areas of oil, gas, power, and the environment. After a grant request is approved, the TDA requests that the government of the host country issue the terms of reference to a short list of American firms. The winner is selected through competitive bidding. Although the bidding process is managed by the relevant entity in the host country, the TDA pays the fees to the selected firm directly.

PSED is an extensive program of assistance aimed at promoting private sector solutions to the energy problems of developing countries. The program concentrates on private power and provides support for studies and training. It finances experts (individuals and firms) to work with host governments to study market demand and prepare the legal and contractual documents for private power investments. It also finances training for local authorities and the expenses of conferences, study tours, and other activities related to cooperation with U.S. industry.

USExim operates as an independent U.S. government agency. Its president and chairman is appointed by the president of the United States. The Board of Directors has five members (president, vice-president, and three others), of whom only three can be from the same political party.

USExim's loans provide fixed-interest-rate financing for U.S. exports. USExim extends direct loans to foreign buyers of U.S. exports and intermediary loans to parties that extend loans to foreign buyers. USExim's rates are in line with OECD guidelines.

USExim guarantees provide repayment protection for private sector loans to creditworthy buyers of exported U.S. goods. USExim may also guarantee payments on cross-border or international leases. Most guarantees provide comprehensive coverage of both political and commercial risks, but guarantees covering political risks only are available.

Overview 6.4 The U.S. Bilateral Assistance Program

- The United States was the largest provider of assistance to developing countries until the early 1990s, but it is now the second-largest provider, after Japan.
- The main agencies are the U.S. Agency for International Development (USAID), U.S. Export-Import Bank (USExim), Overseas Private Investment Corporation (OPIC), and the U.S. Trade and Development Agency (TDA).
- USAID has numerous support programs, many of which cover energy projects, studies, and other types of assistance requirements. A special Private Sector Energy Development Program (PSED) promotes private sector solutions to energy development issues.
- The TDA provides grants for feasibility studies for proposed projects in developing countries.
- USExim provides loans and loan guarantees for procurement of U.S. goods and services.
- OPIC provides loans, guarantees, insurance, and equity to ventures involving significant capital and management participation by American companies.

Of additional relevance to export finance is the Private Export Funding Corporation (PEFCO), which is owned by 43 commercial banks, six industrial companies, and three financial services companies. PEFCO makes fixed-interest-rate loans to foreign buyers of U.S. exports. These loans carry the unconditional guarantee of USExim for timely payment of principal and interest.

OPIC was spun off USAID in 1971. A U.S. government body reporting to the secretary of state, it provides loans, guarantees, insurance, and equity to ventures involving significant capital and management participation by U.S. companies. OPIC's objective is to support projects that improve U.S. global competitiveness, create American jobs, and increase U.S. exports. OPIC's loan and equity facilities are very small. Its guarantee and insurance authorization facility, however, is quite substantial (about $4 billion per year). Guarantees are issued for loans borrowed from U.S. financial institutions or foreign lending institutions that are at least 95 percent U.S. owned. OPIC's insurance covers political risks, including expropriation, political violence, and currency inconvertibility.

Belgium

The Belgian development aid program is channeled through two major institutions: the Belgium Administration for Development Cooperation (BADC) and the Department of Treasury. Most of Belgium's aid is targeted for agriculture and education, particularly in Sub-Saharan Africa.

Responsibility for export credit lies with the Office National du Ducroire (OND), which has some interest in the energy sector, particularly power distribution facilities. The OND provides export credit insurance, and commercial banks provide export financing. Export credits are also provided by Creditexport, a revolving fund financed by private banks and public agencies. The OND's insurance coverage is usually a prerequisite for export credit support from other public and private sources. Copromex, an entity under the authority of the minister of foreign trade, subsidizes interest rates when such action is required to meet competitive offers.

Canada

Canada's development assistance program of about C$2.5 billion annually is channeled through the Canadian International Development Agency (CIDA). CIDA's priority sectors are human resource development, poverty alleviation, and the environment. However, it provides assistance in other sectors for implementation of practical projects that draw on resources of the Canadian private sector. CIDA has been an important source of support for preinvestment work in developing countries. In 1978, it set up the Industrial Cooperation Program to provide financial assistance to Canadian companies seeking business ventures with companies in the developing countries of Africa, Asia, and Latin America. This program provides finances for studies aimed at identification, assessment, and preparation of projects. It also provides finance for studies aimed at improving the technical and institutional efficiency of developing-country companies. The degree of financial support depends on the type of study. The limit is C$500,000 for most studies.

Responsibility for export credit lies with the Export Development Corporation (EDC), which is interested in the oil, gas, and power areas. The EDC provides both financing and insurance to help Canadian exporters compete in other countries. The EDC's loans are generally extended directly to foreign buyers. Loan terms are in line with OECD guidelines. However, the EDC can provide the following additional services:

- Mixed credits. EDC blends export credits at OECD Consensus rates with concessional funds from the Canadian government to produce financing packages competitive with those offered by other exporting credit agencies. Use of this matching program is restricted and depends on the final approval of the government of Canada.
- Parallel financing. For certain countries, Canada offers aid funds through the CIDA, in conjunction with export credits through the ECD, to provide attractive financing packages. This program usually involves the arrangement of separate loan agreements for each type of financing in support of a project.
- Insurance. The ECD offers a wide range of export credit insurance, guarantees, and foreign investment products. This includes coverage against losses relative to bank guarantees and bonds, coverage for equipment, and coverage against losses relating to new investments in foreign countries.

Denmark

Denmark's annual development assistance of about $1.4 billion is channeled through the Danish International Development Assistance (DANIDA). Sectors of concentration include agriculture; social and economic infrastructure; industry; and, on a relatively small basis, energy.

The Industrialization Fund for Developing Countries (IFU) was established in 1967 to support joint ventures between Danish companies and enterprises in developing countries. It has supported several small investments in oil, gas, and power in Latin America.

Export support is provided by the Export Credit Council *(Eksportkreditraadet;* EKR) and the Danish Export Finance Corporation (EFC). The EKR only offers guarantees. The EFC provides loans, with the condition that the EKR provides guarantees. The EKR's guarantee fees depend on the type of coverage (for commercial or political risks) as well as the creditworthiness of the country and the borrowing entity. Interest in the energy sector, for both agencies, is limited to efficiency improvement and promotion of alternative energy.

Finland

Finland's annual development assistance program of $300 to $500 million is implemented by the Department for International Development Cooperation (DIDC), a department within the Ministry of Foreign Affairs (this department was formerly called the Finnish International Development Agency; FINNIDA). Sectors of interest include energy, particularly matters related to the environment. Another institution involved in assisting developing countries is the Finnish Fund for Industrial Development (FINFUND), which finances industrial joint ventures between Finnish enterprises and developing countries.

The export support facilities include two agencies: Finnish Export Credit Limited (FEC) and the Finnish Guarantee Board (FGB). FEC is a joint stock company in which the government of Finland is the majority shareholder. It provides financing for a supplier, buyer, or bank-to-bank credit agreement.

In the FEC lending program, there are three basic schemes (each with different procedures and lending rates): (1) OECD Consensus-based credits, (2) credits at market rates, and (3) concessional or mixed credits for qualified development projects. FEC credits are extended against adequate collateral, such as guarantees by the FGB or guarantees by Finnish or foreign banks acceptable to FEC.

The FGB (also referred to as the VTL) is a government agency supervised by the Ministry of Trade and Industry. The FGB encourages Finnish exports under a comprehensive export credit guarantee program consisting of (1) insurance against political or commercial risks in credit transactions; (2) insurance against political risks connected with investments; and (3) securities for credits granted to exporters, foreign buyers, and financial institutions in connection with financing exports from Finland.

France

France's development assistance program consists of about $6 to $7 billion annually and involves the Ministry of Foreign Affairs, Ministry of Economy and Finance, Ministry of Cooperation, Caisse Française de Developpement (CFD), and Fonds d'Aide et de Cooperation (FAC). Sectors of interest include energy, particularly power facilities. Support is focused on Francophone African countries, although other countries, particularly in the Middle East and Asia, are considered. Almost all of the development assistance is in the form of grants and highly concessional loans.

Policies governing export credit facilities are formulated by two government agencies, the Direction des Relations Economiques Extérieures (DREE) and the French Treasury. In practice, the export support is provided through the Compagnie Française d'Assurance pour le Commerce Extérieur (COFACE), which is France's export credit agency, and the Banque Française du Commerce Extérieur (BFCE). COFACE provides insurance against commercial and political risks. It also insures against political risk on direct investments in developing countries. The BFCE provides financial support on the account of the government by stabilizing interest rates for private export credits.

Germany

The German development assistance program is about $6 to $7 billion annually and involves the following groups:

- The Federal Ministry for Economic Cooperation and Development (*Bundesministerium für Wirtschaftliche Zusammenarbeit*; BMZ), which is responsible for policy, decisionmaking, and funding for the foreign assistance program.
- The Kreditanstalt für Wiederaufbau (KfW), which handles the implementation of the BMZ's financial cooperation program. The KfW also provides mixed financing tied to procurement from German firms. Credits from the export financing program are officially supported and in line with OECD guidelines for official export credits.
- The Technical Assistance Corporation (*Gesellschaft für Technische Zusammenarbeit*; GTZ), which is responsible for implementation of Germany's technical assistance program.
- The Economic Ministry, which is particularly involved in assistance to central and eastern Europe and the FSU.
- The German Investment and Development Company (*Deutsche*

Investitions-und Entwicklunggesellschaft; DEG), which has objectives similar to those of the International Finance Corporation (IFC); that is, it extends loans and makes equity investments in private companies in developing countries to promote the private sector.

All financial and technical assistance is on a grant basis. Energy sector projects, particularly projects that relate to energy-environment matters, are considered. Financing proposals can be submitted by the recipient country, the donor country, or any development institution. However, a request from the would-be recipient country is needed for formal approval. Proposals pass through local German embassies to the Foreign Ministry in Bonn. Local embassy representatives convey proposals to the BMZ, where proposals are evaluated from a development policy standpoint.

The BMZ, as a policymaking body, decides whether a proposal fits into the German aid program. Subsequently, investment and export financing proposals are passed on to the KfW for appraisal, and technical assistance proposals are sent to the GTZ. Both agencies handle implementation once the BMZ issues final approval.

Germany's large export support program involves a number of important players. At the policy level, the Inter-Ministerial Committee is responsible for government policy on export credit support. This committee is composed of representatives from the ministries of economics, finance, foreign affairs, and economic cooperation. Advised by members of the industrial, banking, export, and commercial sectors, it is chaired by the Ministry of Economics.

In the area of implementation, export credit support involves the following agencies:

- Hermes Kreditversicherungs, AG, a private company that insures against credit risks for the account of the government.
- Ausfuhrkredit-Gesellschaft (AKA), an institution owned by a consortium of German commercial banks that provides export credits on market terms and officially supported long-term export credits at preferential rates.
- The KfW, which provides officially supported, long-term export credits at preferential terms (aid credits) and at market terms (acting in this respect as a commercial bank).

Credits financed by special government funds are subject to OECD guidelines. However, the following special services are available:

- Concessional funds can be blended with export credits to help finance German projects in developing countries.
- Untied financial credits (not directly related to German exports) can be extended by the KfW in exceptional cases only for balance-of-payments purposes to ensure raw material supplies.
- Financial assistance is available from KfW or directly from Germany's federal budget to support private investment in developing countries. Federal funds are disbursed through the DEG, which is similar to the IFC. The DEG was established by the German government but operates in line with private sector principles.

Italy

Italy's development assistance program involves about $3 billion annually. The program is implemented by the Directorate General for Development Cooperation (DGCS) of the Ministry of Foreign Affairs. Sectors of interest include power, gas, and petroleum facilities. Aid takes the form of grants and highly concessional loans.

Export credit is administered by the Instituto Centrale per il Credito a Medio Termine (MEDIOCREDITO CENTRALE). Insurance for export credits against political and commercial risks is provided for export credits by the Sezione Speciale per Assicurazione del Credito all' Esportazione (SACE).

Netherlands

The development assistance program involves about $2.5 billion per year and is implemented by the Directorate General for International Cooperation (DGIS), within the Ministry of Foreign Affairs. The program focuses on poverty alleviation but provides some support to the energy sector, particularly in relation to the environment, rural energy matters, and efficiency improvement projects. Grants for technical assistance are available for certain countries.

Commercial banks provide export financing at market rates. Insurance against export credit risks is provided by the Nederlandsche Credietverzekering Maatschapij (NCM), a private company that receives support from the government. An export credit program is administered by the Ministry for Development Cooperation, which in certain cases provides a grant of up to 45 percent of the foreign currency component of a project involving export promotion. The areas of interest for export promotion include power and gas facilities and services.

United Kingdom

The development assistance program consists of about $3 billion annually. The agencies responsible for implementation of the program are as follows:

- The Overseas Development Administration (ODA), which is the primary agency responsible for formulation and implementation of all British development cooperation.
- The Crown Agents, an agency that administers bilateral loans and grants on behalf of the ODA, especially to projects in Africa. For some projects, the Crown Agents also manage procurement, quality control, and transport arrangements on behalf of the recipient country.
- The Commonwealth Development Corporation (CDC), whose board is appointed by the ODA, invests loans and equity in public and private sector projects in developing countries. The CDC obtains its resources by borrowing from the official aid program at concessional rates and from internally generated funds. The CDC invests in independent and integrated power systems in the private sector and in support of privatization. It is also keen to invest in oil and gas businesses in conjunction with the private sector. The CDC is particularly supportive of investments in very poor countries.

The United Kingdom provides development assistance in the form of grants and highly concessional loans. Sectors of interest include energy. Countries of interest are those in the British Commonwealth and Sub-Saharan Africa. Other aid recipients include Indonesia, Egypt, Turkey, Mexico, and countries of eastern Europe and the FSU.

Export credit comes from commercial sources. However, two government agencies are also involved in providing this support:

- The Department of Trade and Industry (DTI), which focuses on commercial and industrial aspects of development cooperation and handles relations with the United Nations Conference on Trade and Development (UNCTAD), and the World Trade Organization. The DTI provides financing in the form of commercial loans (with interest support from the government) and mixed credits—that is, commercial loans with government grants.

- The Export Credit Guarantee Department (ECGD), which is a separate government institution under the Department of Trade and Industry. The ECGD provides insurance against noncommercial risks and subsidizes interest rates on private export credits.

Abu Dhabi

The Abu Dhabi Fund for Arab Economic Development was established in 1971 as an autonomous national development institution belonging to the government of Abu Dhabi. The fund's objective is to assist developing countries by extending project loans, guarantees, technical assistance grants, and equity participation.

Initially, the fund's assistance was restricted to Arab countries. However, in 1974 (as a result of substantially increased resources), the fund's mandate was enlarged to cover all developing countries. Still, 75 percent of total loan commitments is invested in Arab countries; Asian and African countries receive 15 percent and 8 percent, respectively. The largest Arab recipients are Oman (21 percent), Tunisia (8 percent), Egypt (6 percent) and Yemen (5 percent).

By the end of 1995, the Abu Dhabi fund had granted more than $1.6 billion in 120 loans to 48 developing countries. Nearly half of these commitments support extractive and manufacturing industries. Electricity and water supply accounted for nearly 40 percent.

Fund maturities range from 8 to 22 years, depending on the recipient country and the type of project, and include a grace period of 3 to 8 years. The interest rates, including a fee of 0.5 percent, vary from 2 to 6 percent.

Loans are made to governments, companies, or public institutions with government guarantees. Procurement is subject to international bidding procedures.

Australia

The objective of the Australian development cooperation program is to promote sustainable economic growth and social advancement in developing countries. The program is designed to promote Australia's humanitarian concerns, foreign policy, and commercial interests.

Responsibility for Australia's official development assistance program rests with the Minister for Development Cooperation and Pacific Island Affairs, who delegates authority to the Australian Agency for International Development (AusAID).

Annual assistance of about A$1.5 billion covers almost all economic sectors, including mining and energy. Funds are provided in the form of grants. The program focuses on the South Pacific and Southeast Asia region.

Australian development assistance grants may be combined with export credits, provided specific requirements are met. Terms and interest rates are extended in accordance with OECD guidelines. Responsibility for export credit services lies with the Export Finance and Insurance Corporation (EFIC).

EFIC's services include

- Insurance against not receiving payments for exports of goods and services by Australian nationals.
- Direct lending to overseas borrowers for the purchase of Australian capital goods and services.
- Guarantees to lending institutions that finance Australian exporters or make loans to overseas borrowers for the purchase of Australian goods and services.
- Insurance for Australian suppliers of goods and services against unfair calling of performance bonds.
- Insurance of bonds to financiers or overseas buyers on behalf of Australian exporters in support of export contracts.
- Insurance against political risks for Australian firms investing in enterprises overseas.

Austria

Austria's development program assists in utilization of natural resources and energy potential. Responsibility for policy formulation lies with the Federal Chancellery. Allocation of bilateral assistance is carried out by a minister of state who works as part of the Federal Chancellery but implements the program through the Directorate General for Development Cooperation (DGDC).

The program involves about $500 to $600 million per year and focuses on Sub-Saharan Africa, Central America, South Asia, eastern Europe, and the FSU. Assistance is in the form of grants and highly concessional loans.

Responsibility for export credit lies with the Oesterreichische Kontrollbank, AG (OEKB). An interest rate subsidy from the federal budget enables the OEKB to provide concessional export credits under terms consistent with OECD guidelines.

The OEKB places no formal restrictions on sectors or countries of interest. The OEKB provides guarantees for political risks on investments by Austrian companies abroad.

Norway

Norway's development assistance program is aimed at helping all of the world's poorest countries, though most of the aid is concentrated in Africa. Norway's annual contribution of $1.4 billion amounts to 1.1 percent of its GNP, which ranks first in terms of the relative generosity of its development assistance among all OECD countries. Sectors of interest include oil, gas, and power.

Policy aspects of bilateral assistance are handled by the Department of Bilateral Development Cooperation in the Ministry of Foreign Affairs. Operations are handled by the Norwegian Agency for Development Cooperation (NORAD).

Support for exports is provided in the form of loan guarantees by the Garanti-Instituttet for Eksportkreditt (GEIK), a public agency. In addition, NORAD provides export credit in cooperation with Eksportfinans, an export credit agency owned jointly by commercial banks and the GEIK.

Sweden

Sweden bases its development assistance program on a philosophy of solidarity with the underprivileged. It provides about $2 billion per year (about 1 percent of GNP) to the poorest countries in Africa and Asia. Sectors of interest include energy—particularly generation, transmission, and distribution of power.

Before 1995, Sweden had four development cooperation authorities:

- The Swedish International Development Authority (SIDA), responsible for handling most bilateral assistance.
- The Swedish Agency for International Technology and Economic Cooperation (BITS), responsible for promoting economic and social development through transfer of technology. BITS provided grant

financing for technical assistance and concessional credit financing for high-priority investment projects.

- The Swedish Agency for Research Cooperation with Developing Countries (SAREC).
- The Swedish International Enterprise Development Corporation (Swe De Corp), which supported enterprise development through joint-venture investments in developing countries.

In mid-1995, the above agencies merged into SIDA, which is now responsible for all aspects of bilateral cooperation.

Export credits are also provided by SIDA. Guarantees are provided by the Swedish Export Credits Guarantee Board (EKN).

Switzerland

Switzerland's development assistance program involves about $1 billion annually, aimed at low-income developing countries. Sectors of interest are agriculture and social infrastructure, although some assistance is provided to the energy sector. Responsibility for the development assistance program lies primarily with the Directorate for Development Cooperation and Humanitarian Aid (DEH).

Limited concessional funds are made available for export credit, administered through the Federal Office of Foreign Economic Affairs (BAWI).

Spain

Spain's development assistance program consists of about $1 billion annually. Most of this aid has been in the form of balance-of-payments support, debt consolidation, food relief, and technical assistance. Assistance to the energy sector is relatively small. Regions of interest include Central and Latin America, the Middle East, and North Africa.

The responsibility for management and allocation of development aid lies with the Ministry of Economy and Finance (MOEF). For technical cooperation, the responsible institution is the Agency for International Cooperation (AIC).

Export credits are provided (on concessional bases) under the authority of the Secretary of Commerce, through the Institute for External Trade (ICEX). An autonomous institution, the Export Credit Insurance Company (CESCE), provides export credit coverage against political and commercial risks to Spanish enterprises.

Portugal

Portugal's development assistance program involves about $300 million annually and concentrates on countries with which Portugal has historical ties: namely, the Lusophone African countries (called PALOP countries). The sectors of interest are health, education, and agriculture, although limited assistance is occasionally given for energy matters. The primary agencies are

- The Institute for Economic Cooperation (IEC), which coordinates development assistance and defines strategies for providing funds to developing countries.
- The Fund for Economic Cooperation (FCE), which extends grants and loans to Portuguese organizations and enterprises that pursue cooperation in developing countries. Terms are in line with OECD guidelines.

Kuwait

Administered through the Kuwait Fund for Arab Economic Development (Kuwait Fund), the Kuwait aid program is one of the oldest development assistance programs. It was established in 1961 to assist Arab countries and other developing countries by providing financial and technical resources on concessional terms for specific projects. The Kuwait Fund's resources increased substantially after oil prices tripled in 1974. Its mandate was extended at that time to cover all developing countries. Presently, it provides assistance to more than 70 countries in the Middle East, Africa, Asia, eastern Europe, and the FSU.

The volume of development assistance is about $500 million annually. Sectors of interest include energy. The terms of loans depend on the project and per capita income of the recipient country. Some loans include a grant element of up to 75 percent. Interest rates on the loan element range from 0.5 to 5.5 percent annually. All loans and grants are untied. Procurement of goods and services is subject to international bidding procedures. The Kuwait Fund provides grants for preinvestment studies, recruitment of experts, and vocational training.

Responsibility for policy formulation and implementation of the development assistance program lies with the Kuwait Fund, which is directed by a board chaired by the prime minister of Kuwait.

Saudi Arabia

In 1975, the government of Saudi Arabia established the Saudi Fund for

Development, aimed at providing grants and concessional loans to developing countries. Over the next 20 years, the fund provided about $6 billion to low-income countries in Africa and Asia. Sectors of interest include energy, although priority is given to social and physical infrastructure.

The terms of loans vary depending on the country and sector. Some loans include grant elements of up to 80 percent. The interest rate (called service charge) on the loan element varies between 0.5 and 5 percent annually. Loan maturities vary from 16 to 50 years. All loans and grants are untied. Procurement is subject to international competitive bidding procedures.

The Saudi fund is administered by a Board of Directors chaired by the Saudi minister of finance and national economy.

Korea

The Republic of Korea established an Economic Development Cooperation Fund (EDCF) in 1987 to assist developing countries and promote economic cooperation between these countries and Korea. Because the fund is young, the volume of assistance, which has increased annually, is still modest. The fund provided $200 million in assistance in 1995. About half of this amount was allocated to countries in Asia; the rest was allocated to eastern Europe and Africa. About 60 percent of the funds are channeled to developing countries through multilateral institutions; the rest are channeled through direct bilateral relationships.

The bilateral loans are on soft terms. The average annual interest rate is 3.5 percent, and the average maturity is about 21 years, including a 5-year grace period. Sectors of interest include energy—particularly power transmission, which is an area in which Korean manufacturers are eager to export.

Agencies involved in bilateral assistance are

- The Ministry of Finance and Economy, which plays a key role in shaping the overall assistance program.
- The Ministry of Foreign Affairs, which receives requests for assistance through a worldwide network of embassies.
- The Export-Import Bank of Korea (KExim), which is the main government body responsible for issuing bilateral loans.
- The Korea International Cooperation Agency (KOICA), which handles bilateral grants, including grants for technical cooperation.

7

ACCESSING COMMERCIAL FUNDS

During the 1980s and 1990s, many instruments appeared to channel commercial funds to projects in various parts of the world. Corporations in industrialized countries increasingly use these instruments to raise equity capital and debt financing. Companies and projects in the developing world have less diverse options; however, significant new tools are available.

Commercial funds for energy companies in developing countries come from six major sources:

- Domestic capital markets, which provide both equity capital and debt financing.
- International commercial banks, from which companies can get independent loans from several banks or syndicated loans (provided by a number of banks on a pro rata basis under identical terms and conditions).
- International equity markets, where a company's equity shares can be sold through public offerings or private placement with sophisticated institutional investors (pension funds, insurance companies, mutual funds, and so on).
- International bond markets, where a company can borrow funds by issuing bonds through public offering or private placement with commercial banks and institutional investors.
- Specialized energy funds. These are of two distinct types. First, several energy funds have been created with partial government sponsorship to facilitate private sector investment in the energy sectors of particular countries. Second, several funds have been created purely on a commercial basis to channel funds from investors in the world's major financial centers for lending and equity investment in developing countries.

- Financial contributions from project beneficiaries. These contributors include equipment suppliers and contractors (who may provide equity or debt finance in order to win the business of building the project facilities), sellers of fuel to the project plant, and purchasers of the project output (who may make financial contributions because they want to do business with the project company).

Overview 7.1 Major Sources of Commercial Finance

The major sources of commercial finance for energy projects in developing countries are

- Domestic capital markets.
- International commercial banks.
- International equity markets.
- Bond markets in the United States, Europe, and Japan.
- Specialized energy funds.
- Equity and debt finance contributions from project beneficiaries.

DOMESTIC CAPITAL MARKETS

The development of domestic capital markets has been a relatively recent event for most developing countries. For decades, governments of many countries did not support such markets, primarily because these markets can decrease government control over national savings. Structural changes of the 1990s convinced many developing-country governments that one of the main pillars of a competitive economy is an efficient capital market. Now, many countries promote their capital markets. The outcomes of these changes differ by country depending on the stage of economic development and other aspects of the business environment.

Although financing energy projects through domestic capital markets has been limited, recent developments in several countries (Thailand, Malaysia, Philippines, Chile, India, Brazil, and others) have shown that domestic capital markets can rapidly become important sources of finance for energy projects, particularly gas and power projects. Between 1990 and

1995, the domestic capital markets of developing countries in East Asia doubled in size. These markets are expected to triple between 1995 and 2005.

Energy projects can access domestic savings through several instruments. First, to mobilize equity, a project company can issue shares on the domestic stock market. This is normally feasible for well-established companies already engaged in the sector. A new project company would have difficulty raising funds on the stock market, however. Such a company could approach institutional investors (such as pension funds and insurance companies), which are better able to assess potential project risks and rewards and may be willing to place private equity with the project company. In some developing countries, wealthy individuals are also potential contributors of equity for a new plant. Second, to mobilize debt financing, a project company can borrow from local commercial banks or issue local bonds. Local commercial banks are not normally able or willing to take the risks perceived as part of a large project. However, they are sometimes tempted to lend when they see reputable international financiers involved in a project. Finally, a project company can issue bonds, although bond purchasers are very risk-averse and are likely to purchase bonds only if the company is well established in the country.

Overview 7.2 The Role of Domestic Capital Markets

Domestic capital markets have developed rapidly in many developing countries. Energy projects can raise equity from domestic institutional investors, private investors, and domestic stock markets. Debt financing can be mobilized from local commercial banks or by issuing local bonds through well-established companies or behind reputable international financiers.

INTERNATIONAL COMMERCIAL BANKS

In the first half of the 1980s, commercial banks represented a major source of financing for energy projects in developing countries. During this period, commercial banks accepted highly leveraged debt transactions, which often gave them little return and left them with high-risk portfolios. The debt crises of the mid-1980s resulted in substantial losses (which continued to be written off for a number of years thereafter) to the commercial banks. Since then, commercial banks have become far more selective about which projects to support. Banks now place greater emphasis on the development of a strong security package and sharing of risks with other financiers. As a result, most loans to large projects are syndicated. A syndicated loan is provided by a number of banks (on a pro rata basis) under identical terms and conditions. These loans carry floating rates, which are normally based on a fixed spread over an index such as the London Interbank Offer Rate (LIBOR). Although the interest rate of syndicated loans is always floating, the loans can be structured to give maximum flexibility by including

- Floating-rate/fixed-rate tranches, whereby borrowers receive the option to switch part or all of the loan into fixed-rate tranches.
- Multicurrency options, where borrowers receive the option to switch part or all of the loan into any convertible currency.
- Longer maturities, which can be as long as 10 to 15 years.

Loan syndication has become similar to issuing bonds. However, specific features make syndication more suitable for the debt financing of energy projects. First, project sponsors can mobilize large loans through loan syndication. Second, commercial banks that participate in loan syndication are usually sophisticated and familiar with the complex credit risks that may be present in financing a project. Third, costs (including a management facility fee) are less than those involved in issuing bonds.

The energy companies of developing countries often access the syndicated debt market through multilateral agencies. The International Finance Corporation (IFC) is well recognized for its ability to syndicate loans for such companies. The IFC's participation provides substantial comfort for commercial banks to participate in debt financing because, as Lender of Record, the IFC (1) handles all the technical and legal matters including risk mitigation measures; (2) ensures that a default on repayment to any of the participating banks will be viewed as a default on the entire syndicated loan and, thereby, a default on repayment of the IFC loan; and (3) is responsible for administration of the syndicated loan and all matters related to the disbursement and repayment of the loans. The IFC also acts as the lead manager lender and syndicator. Other multilateral institutions—the World Bank, Asian Development Bank, European Bank for

Reconstruction and Development, and Inter-American Development Bank—may provide partial risk guarantees to commercial banks for lending to energy companies in developing countries.

Overview 7.3
The Role of International Commercial Banks

International commercial banks are now more cautious about lending to projects in developing countries than they were in the 1970s and early 1980s. However, they are still a major source of finance for energy projects. Their participation is increasingly in the form of syndicated loans, which are provided by a number of banks. The International Finance Corporation and some other multilaterals play important roles in structuring syndicated loans for developing countries.

RAISING EQUITY ON INTERNATIONAL CAPITAL MARKETS

The globalization of capital markets has provided an opportunity for raising equity on international markets. Oil, gas, and power companies of industrialized countries use these markets on a regular basis. However, companies of developing countries have comparatively limited access to these markets because

- Some industrialized countries impose legal restrictions on the equity participation of their institutional investors in projects located in developing countries.
- Reliable information on the relevant performance of entities in developing countries is often seriously lacking. Without such information, investors cannot make reliable judgments about projects in these countries and therefore are not willing to risk their capital in such projects.

In the early 1990s, changes in the regulatory and supervisory structures of the security and exchange operations of industrialized countries helped companies of developing countries mobilize equity. For example, in the United States, public offerings of equity shares normally require registration with the U.S. Securities and Exchange Commission (SEC), which in turn requires extensive, detailed information about the company. However, an SEC ruling in April 1990 (known as Rule 144A) allows qualified institutional investors to buy restricted securities not registered with the SEC. These securities can be traded freely by the public after three years of trade under 144A. This rule has enabled some power and oil companies of developing countries to raise equity on the U.S. market using American Depository Receipts (ADR). ADRs enable foreign companies to issue equity on the U.S. market without complex settlement and transfer mechanisms. They are issued by a U.S. depository bank, but the underlying shares of the company are held by a custodian bank in the home country for the depository bank. ADRs can be traded in the United States on a recognized national exchange or placed privately with institutional investors under Rule 144A. In the case of energy companies of developing countries, ADRs are mostly placed privately with institutional investors. Equity placements, whether public or private, benefit substantially from some type of underwriting by a credible institution, such as the IFC.

Overview 7.4
Raising Equity on the International Capital Markets

Equity can be raised on international capital markets by public offering of the shares of the company or through private placement of these shares with institutional investors. Public offering of equity shares requires approval from relevant authorities. Private placements are comparatively much less restrictive.

INTERNATIONAL BOND MARKETS

The major bond markets are in the United States, Japan, Germany, and the United Kingdom, although other bond markets are growing rapidly in Europe and Asia. Traditionally, these major markets have been used by domestic borrowers, including each country's federal and local governments, private corporations, and commercial banks. Foreign bonds—that is, bonds issued by a foreign borrower—have also been in the market for a long time but expanded in volume only in the 1980s and 1990s.

The relevance of the major bond markets to financing of energy projects in developing countries is twofold. First, many parent companies (for example, international oil companies) are considered domestic borrowers in the major bond markets. They issue corporate bonds and use the borrowed funds for equity or debt financing for their projects in developing countries. Second, energy companies of developing countries issue bonds in the major bond markets. Their issues are considered foreign bonds and are subject to foreign bond regulations.

Another important instrument for funding energy projects in developing countries on international bond markets is the *Eurobond.* The origin of the Eurobond market goes back to 1963, after an "interest equalization tax" was levied on purchases of foreign securities by U.S. nationals. This tax made foreign bond issues less attractive and effectively closed New York to foreign borrowers. As a result, the market moved to London, where borrowers were able to tap the *Eurocurrency* market. Today, the prefix *Euro* does not mean the bond must be issued in a European market or in a European currency. A Eurobond can be issued in any convertible currency (*Eurodollar* bonds, *Euroyen* bonds, *Eurosterling* bonds, and so on). However, a Eurobond is underwritten by international syndicate of banks and other security firms and is sold exclusively in countries other than the country in whose currency the issue is denominated. For example, a Eurodollar bond can be issued by U.S. or non-U.S. corporations and sold to investors internationally. The main incentives for issuing Eurobonds are the absence of regulatory interference, less-stringent disclosure requirements, and favorable tax treatment. Most governments impose tight controls on issuers of securities denominated in their currency and sold within their national boundaries, but they have less-stringent limitations for securities denominated in foreign currencies and sold within their markets to foreign currencies holders. In effect, Eurobond sales fall outside of the regulatory domain of any single nation.

The U.S. bond market is the largest in terms of both its volume and ratio of volume to the country's GDP. It has the most developed and sophisticated issuing techniques (transaction transparency, bond diversity, and regulatory mechanism). Bonds can be sold to the public or placed privately with institutional investors (pension funds, insurance companies, mutual funds, and so on). Publicly offered bonds require registration with the SEC and rating of the borrower (or the issue) by a rating agency such as Standard & Poor's or Moody's. Bonds can be placed privately without SEC registration or a formal rating.

In terms of market participants, the U.S. bond market is dominated by government securities. The federal government, municipalities, and local governments finance a large portion of their debt by selling bonds to the public. The corporate sector, which finances a substantial portion of investments by issuing bonds, is second to government agencies in terms of volume of bonds issued.

The corporate bond market generally encompasses two types of bonds: those issued by corporations and those issued by banks. Bank bonds are an important market component in Germany and Japan but not in the United States. This difference reflects the role that the Japanese and German banks have as the main institutions that finance private sector investment.

During the 1980s, the U.S. bond market became more hospitable to foreign bonds (also called Yankee bonds), which have made rapid strides into the United States since then. A major factor working in favor of Yankee bonds has been SEC Rule 144A, under which qualified institutional buyers that manage at least $100 million in securities of nonaffiliated companies on a discretionary basis may sell securities to each other without formal registration with the SEC. Furthermore, in 1994, the SEC streamlined registration and reporting requirements for foreign companies.

The Japanese bond market is the second-largest bond market in the world (after the United States). Government bonds represent more than half of the total volume. The government role in the bond market is important because the government is the largest borrower and has played a major role in shaping the market. Commercial banks, which issue bonds to supplement deposits for relending to corporations, are second to government agencies. Since the 1970s, Japan has removed most impediments to entry of foreign corporations into its bond market. There are two types of yen-denominated foreign bonds: Samurai bonds and Shibosai bonds. Samurai bonds are publicly offered issues, mainly used by large, creditworthy lenders such as supranational institutions (for example, the World Bank and the Asian Development Bank) and sovereign governments. Shibosai bonds are privately placed bonds issued by foreigners.

The German bond market grew very rapidly in the 1980s. The major forces behind this growth were the government, which borrows substantially on the bond market, and commercial banks, which use the bond market to supplement their deposits for lending to the private sector. The procedures for issuing bank bonds are less complicated than those for other corporate bonds. Issuers of bank bonds are normally assigned the special status of "frequent issuers" under German banking law. This enables them to issue a number of bonds based on a general prospectus (usually published every three years) and a very short information prospectus for individual bonds.

The lead manager of a deutschemark-denominated domestic bond issue must be a bank incorporated in Germany, and the bond issue must normally be floated in Germany. Every lead manager has to inform the Bundesbank of the issue, amount, terms, and placing method. Public bonds have to be listed on a German exchange.

Issuance of foreign bonds on the German market has expanded in the 1990s because of certain simplifications (for example, the abolition of the compulsory period of notice to the Bundesbank for new issues) and changes in tax laws and regulations. Foreign issuers of bonds are mainly industrial firms and commercial banks.

The growth of the U.K. bond market has been limited. In terms of both volume and ratio of volume to GDP, the U.K. bond market is far smaller than those of the United States, Japan, or Germany. However, within the U.K. bond market, the share of international bonds is significant.

Although London is still a major center for selling Eurobonds (particularly Eurodollar bonds), Eurobonds are increasingly sold on other markets. On the Japanese bond market non-yen-denominated bonds are sold publicly (called Shogun bonds) and placed privately (called Geisha bonds). The yen-denominated Eurobonds (called Daimyo bonds) were introduced in 1987 and are listed on the Luxembourg exchange.

The Eurobond market attracts prestigious borrowers from both the public and private sectors. Because of the character of this international market, only substantial borrowers (usually those whose household names are well known to investors) can tap the market. It is rare to find issues secured by assets. Furthermore, evidence of the high quality of borrowers can be found in the low number of defaults. Eurobonds' maturities normally range from 2 to 10 years, and normal sizes vary from $100 million to $3 billion. Interest is paid on a fixed- or floating-rate basis and is paid free of withholding tax. The cost of borrowing varies according to credit quality and maturity. Borrowers require no prior registration or approval from any governmental authority, and many issuers enter the market quickly once market conditions become favorable.

Overview 7.5
Borrowing on the International Bond Markets

- The major bond markets are those of the United States, Japan, and Europe. These markets are primarily used by domestic borrowers, including the host country's government, private corporations, and banks.
- Issuance of foreign bonds on each of these markets is of recent origin but has expanded rapidly since the early 1980s.
- Another important instrument for international borrowing is the Eurobond, which is underwritten by an international syndicate of banks and other security firms and sold exclusively in countries other than the country in whose currency the issue is denominated.
- Borrowing on the major bond markets can be undertaken through public issuance or private placement of bonds. Private placement of bonds is much less regulated.
- Since the early 1990s, energy companies of developing countries have tapped these markets by private placement and, in a few cases, by public issuance of bonds.

Since the early 1990s, a large number of energy companies of developing countries have borrowed on major bond markets. Although most of this borrowing has been through private placement on the U.S. 144A market, some issues have utilized the U.S. public bond market, the Japanese private and public market, and the Eurobond market. For example, Mexico's national petroleum company, Pemex, successfully used the U.S. 144A market in the early 1990s. In 1994, Pemex took a further step by issuing a publicly registered Yankee bond in the amount of $200 million and a Samurai bond in the yen equivalent of $200 million. In the same year, the Argentine Yacimientos Petroliferos Fiscales (YPF) issued a $350 million, 10-year Yankee bond that sold easily.

Borrowing on foreign and international bond markets by energy companies of developing countries is facilitated when the bond issue is supported by a multilateral institution. For example, the National Power Corporation (NPC) of the Philippines issued a 15-year, $100 million bond in 1994. The issue was guaranteed by the World Bank and placed in the U.S. 144A and Eurobond markets. The World Bank support was structured as a *put option* for principal repayment at maturity. (The put option allows bond-holders to present or "put" their bonds to the World Bank at maturity for repayment of principal.) The World Bank guarantee is believed to have stretched the maturity, facilitated market access, and reduced the cost of borrowing. Before this issue, 10 years was the longest maturity attained by any issue from a Philippine sovereign entity. In addition, the cost of the NPC issue was 2.5 percent over the yield of U.S. Treasuries of the same maturity, which compares favorably with the cost of previous issues. Other examples of borrowing on major bond markets are discussed in chapter 10, which presents financial structures for several energy projects in developing countries.

SPECIALIZED ENERGY FUNDS

Specialized energy funds are aimed at mobilizing private sector resources. They fall into two general categories: government-sponsored funds and private funds.

Government-sponsored funds encourage private sector participation in the energy sector by providing long-term loans to private sponsors or loan guarantees to commercial lenders willing to lend to private sponsors. Examples include the energy funds of Pakistan and Jamaica. All such funds were created through the support of corresponding governments and with assistance from multilateral and bilateral agencies. The Pakistan fund provides loans for any private project in the energy sector.

The Jamaican fund was specifically formed to provide loans to two private power projects. Funding is provided by the World Bank and the Inter-American Development Bank. The loan maturities are up to 23 years with yearly adjustable interest rates based on U.S. Treasury notes plus a spread. The size of the Pakistan fund is $400 million, of which $150 million is provided by the World Bank and the rest by the Japanese Export-Import Bank, U.S. Agency for International Development, U.S. Export-Import Bank, and aid agencies of Italy and France.

Private funds are formed on a commercial basis with the objective of making aggressive returns either through equity participation or lending to private companies in developing countries. Capital is contributed by institutional investors and large companies. For institutional investors, these funds serve as risk mitigators and facilitators of participation in projects in developing countries. For example, a pension fund in Seattle may not wish to invest directly in a power plant in Thailand because this may seem too risky. However, the pension fund may feel comfortable investing in a global power fund that in turn invests in a diversified portfolio of power projects in several countries, including Thailand. Participation in private energy funds may also be viewed as an opportunity for market development for certain suppliers of goods and services such as power equipment manufacturers and engineering firms.

Private funds are sector- or region-specific. For example, the Scudder Latin American Trust for Independent Power, formed in 1993 with the assistance of the IFC, provides equity for private power projects in Latin America and the Caribbean. The initial resources of about $100 million were provided by four lead investors, and further resources are mobilized from institutional investors on a private placement basis.

Overview 7.6 Specialized Energy Funds

Specialized energy funds are of two types:

- Government-sponsored funds, such as the energy funds of Pakistan and Jamaica.
- Private funds, such as the Scudder Latin American Trust for Independent Power and the Global Power Fund.

The Global Power Fund, formed in 1994, received initial capital of $250 million from its three shareholders. It provides a mix of financing, including equity, subordinated debt, completion guarantees, and bridge financing to cover the construction risks of power projects.

FINANCIAL CONTRIBUTIONS FROM PROJECT BENEFICIARIES

Some project beneficiaries are willing and able to make equity or debt-financing contributions because they are interested in providing inputs to the plant or purchasing its output. Equipment suppliers and contractors represent the most important source in this category. Most manufacturers of large capital equipment have set up captive financing companies for their products. Usually these companies are specifically designed to generate incremental sales by dealing with less-than-investment-grade credits. Competition often forces such companies to offer very competitive rates not otherwise available on the market. Suppliers of equipment and services can also play an important role in mobilizing funds from the bilateral agencies of their own countries (see chapter 6).

THE ROLE OF CREDIT RATING AGENCIES

Several credit agencies, including Standard & Poor's Rating Group (S&P's), Moody's, and Duff & Phelps, evaluate the default risks of lending to various companies and countries around the world. These ratings significantly affect the costs and availability of debt financing.

The private capital markets (for example, commercial banks and private lenders) normally ensure debt repayment by strict control through loan covenants, terms, and other conditions. However, raising funds on the public and quasi-public capital markets depends on establishing the borrower's inherent creditworthiness in terms of its ability to assure the lender of timely debt service. This notion of control versus credit (that is, recognizing the lack of any direct negotiation when dealing with investors in the public capital markets) is a fundamental distinction between the private and public debt markets. In the public debt market, the lenders are concerned only with timeliness of repayment because they have neither the ability nor the willingness to become entangled in a dispute. Thus, ratings are normally based on the assumption that lenders remain passive and will not be required to get actively involved at any time.

Ratings are available for debt issues by a large number of corporations as well as for project-specific financing. In addition, *sovereign ratings* are provided to measure the ability and willingness of a country to service its debts.

Assessment and assignment of debt ratings to corporate and municipal bonds represent the traditional area of activity of the rating agencies.

Credit ratings are carried out for each debt issue and take into account the borrower's default risk, the nature of the debt obligation, the protection that the issue affords, and its relative position in bankruptcy.

Rating of project-specific financing is relatively new but growing rapidly. This type of rating focuses on the ability of a financing entity, whether a developer or a special-purpose vehicle, to make timely payment of principal and interest to bondholders. For example, in rating the financing of an independent power project, S&P's examines transparency of the regulatory environment, contracts governing construction of the project, fuel and operation and maintenance costs, and sales of output. It also considers the creditworthiness of the project sponsors and assesses financial performance across a range of scenarios. Although the evaluation encompasses numerous aspects of risk mitigation, project management, and financial performance, the bottom line will be to assess prospects for timely debt repayment.

Overview 7.7 The Role of Rating Agencies

- Rating agencies, such as Standard & Poor's, Moody's, and Duff & Phelps, play an important role by evaluating the creditworthiness of countries, companies, and bond issues.
- The rating assesses the ability and willingness for timely repayment of a debt.
- The main users of these rating are public investors and lenders. However, most parties that deal with projects, companies, or developing countries look at the corresponding rating or ratings if available.

Sovereign ratings measure the ability and willingness of a country to service its debts. S&P's sovereign analysis looks at two basic elements—political risks and economic risks. In assessing political risks, S&P's considers the stability of the political system, the social environment, and international relations. Orderly succession in political leadership, the system's adaptability to changing circumstances, and the extent of social consensus on basic questions contribute to a strong sovereign rating. The degree of integration into multilateral trade and international financial systems is further regarded as an incentive for a government to honor its foreign obligations.

S&P's economic analysis looks at a government's external financial position, balance-of-payments flexibility, and management of the economy as well as the country's overall economic prospects. The primary aim of this analysis is to assess the debt burden carried by the country and the likely evolution of the country's debt-servicing capacity.

An analytical use of sovereign ratings is reflected in S&P's sovereign ceiling ratings criterion, which states that no entity domiciled in a country can be rated higher than the country itself. This policy recognizes the central government's control over fiscal and monetary policies, foreign exchange controls, and regulation, which collectively render the government's credit standing superior to that of any other debtor in the nation.

Interpretation of Ratings

S&P's rating spectrum runs from AAA to DDD and carries the following interpretation:

- AAA bonds are judged to be of the best quality. These "gilt edge" bonds carry the smallest degree of investment risk. Interest payments are protected by a large or exceptionally stable margin, and principal is secure. Although the various protective elements are likely to change, changes are unlikely to impair the fundamentally strong position of such issues.
- AA bonds are judged to be of high quality by all standards. Together with the AAA group, they comprise what are generally known as high-grade bonds. They are rated lower than the best bonds because margins of protection may not be as large as in AAA securities, fluctuation of protective elements may be of greater amplitude, or other elements may be present that make the long-term risks appear somewhat larger than in AAA securities.
- A bonds possess many favorable investment attributes and are considered upper-medium-grade obligations. Factors giving security to principal and interest are considered adequate, but elements may be present that suggest susceptibility to impairment in the future.
- BAA bonds are considered medium-grade obligations; that is, they are neither highly protected nor poorly secured. Interest payments and principal security appear adequate for the present, but certain protective

elements may be lacking or may be characteristically unreliable over any great length of time. Such bonds lack outstanding investment characteristics and in fact have speculative characteristics as well.

- BA bonds are judged to have speculative elements; their future cannot be considered well assured. Often the protection of interest and principal payments may be very moderate and thereby not well safeguarded during both good and bad times over the future. Uncertainty of position characterizes bonds in this class.
- B bonds generally lack characteristics of a desirable investment. Assurance of interest and principal payments or of maintenance of other terms of the contract over any long period may be small.
- CAA bonds are of poor standing. Such issues may be in default, or elements of danger may be present with respect to principal or interest.
- CA bonds represent obligations speculative in a high degree. Such issues are often in default or have other marked shortcomings.
- C bonds are the lowest-rated class of bonds, and issues so rated can be regarded as having extremely poor prospects of ever attaining any real investment standing.
- D bonds are an issue that is already in payment default or whose issuer has filed for bankruptcy.

Moody's rating system has similar interpretations. The correspondence between Moody's and S&P's ratings is as follows:

S&P	**Moody's**
AAA	Aaa
AA	Aa
A	A
BBB	Baa
BB	Ba
B	B
CCC	Caa
CC	Ca
C	C
D	D

PART III

DESIGNING AN ACCEPTABLE PROJECT PACKAGE

The ultimate objective of the project package is to convince investors and financiers that funds put into the proposed project result in a sufficiently high and secure return. Thus, the two intertwined questions—the project return (or, for senior lenders, cash flow) and its corresponding risks—

represent the cornerstone of project preparation, around which knowledge about several other areas is built. These other areas include technical soundness of the project, environmental concerns, market potential, and political environment and are investigated to ensure that the project return remains adequate, stable, and safe (most major lenders have declared that they consider sustainable development and environmental management independently and in addition to adequate and safe return).

Although many energy projects are built to serve the domestic market, they are always linked to international markets. The general framework for analyzing the viability of a project involves the following steps:

- Assessing potential demand for project output by forecasting consumption by various customer groups and analyzing existing and forthcoming supply capacity. In the event that part or all of the project output is aimed at exports, the analysis should cover the specific markets targeted by the project.
- In relation to the above step, the evolution of energy prices, the tariff structure, and all relevant laws and regulations should be analyzed to arrive at a realistic projection of future prices for output of the energy project.
- International norms and market conditions should be reviewed to establish appropriate references for capital and operation and maintenance (O&M) costs. These could include average prices of capital equipment, fuel prices, and interest rates.
- Local conditions affecting capital and O&M costs should be analyzed, and adjustments should be made according to international average costs. These could include a national "country multiplier," physical conditions (for example, plant site, conditions and remoteness, and pipeline terrain); environmental constraints; laws and regulations; and local input conditions (for example, access and communications infrastructure, fuel and labor markets).

There are many interactions among the above four steps. However, the overall aim of the analysis is to use results of the first two steps to arrive at the likely stream of project revenues, as well as possible variations in these revenues. In the same way, results of the third and fourth steps establish a stream of expected project costs, as well as possible variations in these costs. The revenue and cost streams, along with the range of their variations, form the basis of the risk-reward profile of investment in the proposed project.

If the resultant risk-reward profile is attractive, then the ownership and financing structures are designed based on the analysis of availability of funds from public/private, domestic/foreign, and international sources.

A systematic approach to project preparation involves

- Analyzing the international and local business environment.
- Establishing project viability.
- Structuring the financing package.

Accordingly, Part III of this book contains a chapter on each topic. In addition, because environmental concerns have become very important in the preparation of energy projects, a chapter is devoted to this matter. Also of practical use is Appendix B, which contains a brief description of various sources of information that can be utilized in preparing energy projects.

8

Analyzing the Business Environment

Analysis of the business environment provides a reasonable assessment of the chances that a proposed project will achieve its economic and financial objectives. In practice, this assessment boils down to establishing a basis for assumptions made to construct streams of project costs and benefits.

Although various elements of the business environment are intertwined, the analysis normally focuses on three general categories:

- The project-specific business environment.
- The host country's business environment.
- The international business environment.

THE PROJECT

The project-related analysis aims at providing an overview of the institutional, technical, and cost aspects of the project. The institutional aspects focus on

- Company establishment and ownership—for example, the background, purpose, and shareholdership.
- Organizations and management—for example, composition of the board of directors, relationship with the chief executive officer, and company organization.
- Employment and staffing—for example, the acquisition of expertise, employment of local staff, and promotion of indigenous expertise.
- Background information about project sponsors—for example, the ownership structure, business objective, management, capital, and financial and operating performance of each sponsor.

Figure 8.1
Major Interactions Among the Project Environment and the Business Environment

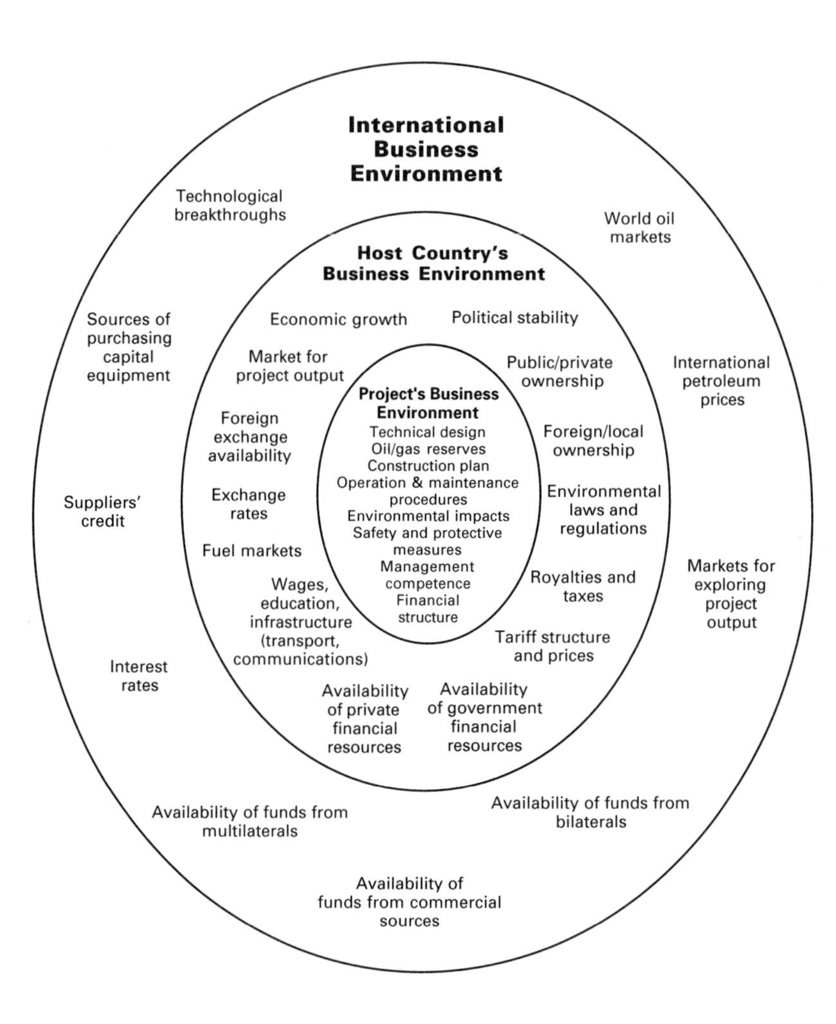

The technical analysis of the project covers all important issues related to design, construction, and operation of the project. The major items are

- The project description, including assurances that the design, engineering, procurement, and construction of project facilities will be in accordance with sound professional engineering practices and based on company guidelines in conjunction with relevant applicable international codes and standards and local legal and regulatory limitations. Strong performance and reliability "completion" tests are essential ingredients.
- The project site, including description of access to transportation facilities, water and wastewater facilities, human habitation, fauna, and flora, which are important in determining project costs and environment impacts.
- The plant operating philosophy and procedures, including the degree of safety, reliability, flexibility, and efficiency. Safety is a major concern in all energy projects. Accordingly, this area receives great emphasis at the stages of design and construction. Nevertheless, minimizing risks also requires effective working and operating procedures, use of preventative maintenance practices, strict adherence to instructions and guidelines, and adequate incentives to facilitate these procedures. Reliability, flexibility, and efficiency are largely related to project design but can be enhanced through effective operating procedures.
- Facilities maintenance, including assurance of regular maintenance as well as plans for shutdowns for major scheduled maintenance.
- Safety and protective measures. These are discussed extensively in the chapter on the environmental assessment of the project (chapter 11). The issues related to design and operation of the plant are summarized here.
- Auxiliary services, including electric power, air and water supply, cooling water systems, and fuel systems needed to support the main plant or facility. The design features and operations standards of each facility should be explained sufficiently.
- Technical assistance agreements and arrangements to ensure that required expertise is available during project construction and operation. In power projects, technical assistance is provided through international consultants or advisors. In oil projects, foreign partners provide substantial assistance. The project document should include terms of reference for all technical assistance services, particularly if a new company is to undertake the project.

The final discussion in the project section should cover expected project costs, procurement, and implementation arrangements. Project costs are normally estimated based on previous experiences in the country, similar projects in other countries, and conditions related to location of the proposed project. In addition, when the project may be implemented based on a turnkey contract, formal and informal quotations may assist in refining cost estimates.

Although project costs are definitely dependent on time, location, and business environment, some average cost numbers have emerged from international experience. These rule-of-thumb figures are often used by experts to arrive at the first-cut estimates for each type of energy project. These "U.S. Gulf" estimates must then be adjusted by a country multiplier. The multiplier varies between 1.1 and 1.25 depending on the country and the region.

In order to link costs to financing needs, it is necessary to separate the costs into foreign and local currency. Normally, local partners, including state entities, feel most comfortable taking responsibility for providing the local currency component of the project cost while leaving the rest for foreign equity and debt financing It is also necessary to separate the costs by items that can be procured independently. The actual procurement arrangement cannot be finalized until the financing package is structured. Indeed, procurement for each project component is often chosen or modified in accordance with requirements of the agency or company providing finance for acquisition of that component. A normal practice is to acquire the main equipment from countries that have substantial suppliers or provide bilateral credit. The procurement arrangement is then made according to requirements of the provider of the credit.

The project construction schedule should include completion dates for the main project components as well as for the auxiliary systems, input deliveries, and offtake and marketing of the output. In an attempt to show that the schedule is realistic, project sponsors refer to their experience in the country. If experience is lacking, they can refer to experience in other countries as well as to other people's experience in the host country. Again, if a turnkey contract is planned, completion dates should have been discussed tentatively with prospective contractors.

HOST COUNTRY'S BUSINESS ENVIRONMENT

Analysis of the host country's business environment should be kept well focused to provide useful analysis of the project's costs and benefits, based

Table 8.1 Rule-of-Thumb Average Cost Figures

- Upstream development costs vary substantially depending on field characteristics and the water depth, if offshore. Consequently, no generally accepted average cost figures are available.
- Onshore pipeline construction costs vary between $15 and $30 per inch-diameter per meter, depending on terrain and other characteristics. A figure of $20 per inch per meter is normally used for first-cut estimates of pipeline investments inclusive of material and construction costs. Offshore pipelines may cost up to 100 per cent more than onshore pipelines.
- The average cost of transporting gas depends on the volume as follows:

Volume (MMCFD)	Average Cost ($ per MMBTU per 1,000 km)
1,000	0.90
2,000	0.70
3,000 and above	0.50

Offshore transportation may cost up to twice the above numbers.

- Refinery construction costs depend on the degree of technology sophistication and operational flexibility. For simple hydroskimming refineries, average costs are roughly estimated as follows:

Capacity (b/d)	25,000	60,000	120,000	200,000
Capital cost ($/bl of capacity)				
Refinery without infrastructure improvement	3,200	2,500	2,000	1,850
Refinery with infrastructure improvement	4,000	3,250	2,500	2,250
Greenfield refinery	5,600	4,000	3,500	3,200

- Construction cost of power plants is estimated at
  - $1,100/kW for coal-based power plants, including flue-gas desulfurization (FGD) facilities.
  - $900/kW for oil-fired steam plants.
  - $650/kW for gas-based combined-cycle plants.
- Investment costs of power transmission and distribution vary substantially depending on population density and stage of the development of the existing system. Some rough approximations can be made based on $125 to $150/kw for transmission and $175 to $200/kw for distribution facilities.
- All of the cost figures above are in 1995 dollars. A country multiplier between 1.1 and 1.25 is applied depending on the country and the region.

Figure 8.2 Factors Affecting Project Viability

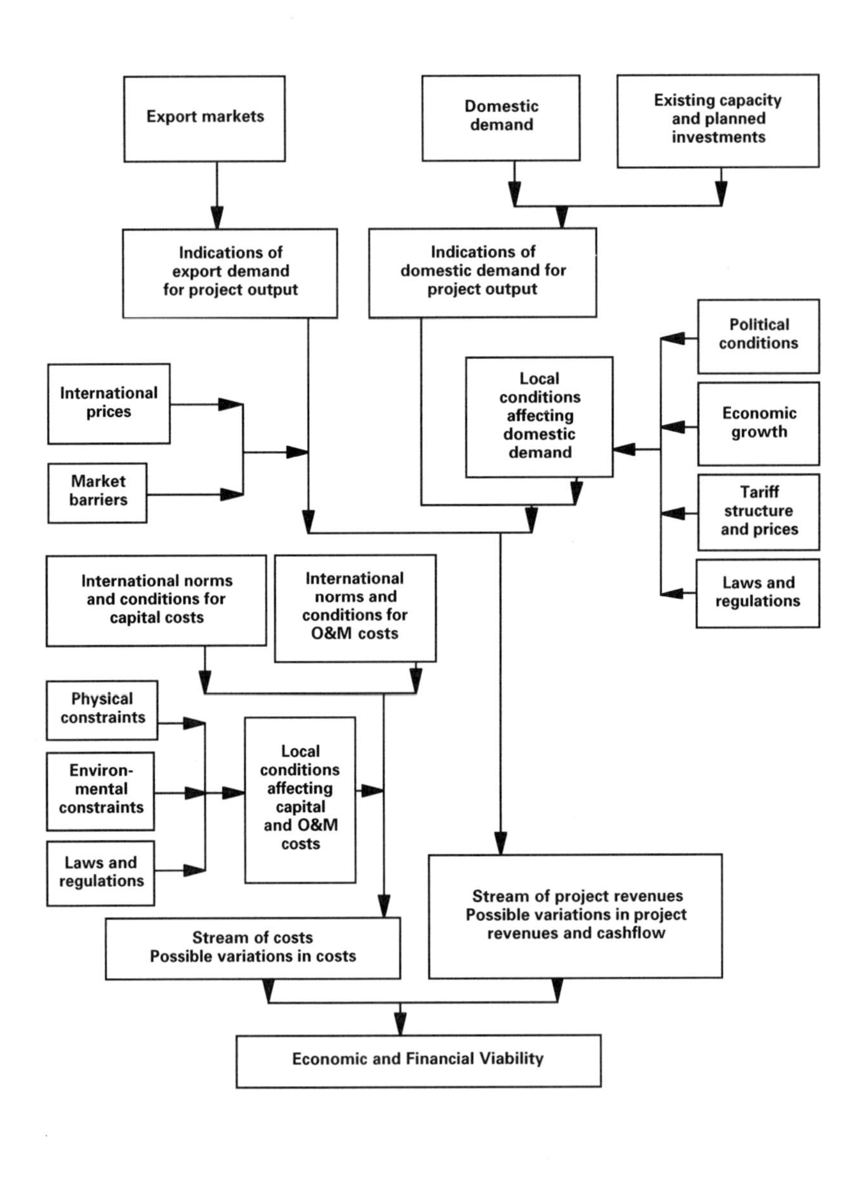

on the current stakes and reasonably foreseeable developments. The major areas to be addressed are

- The political system and circumstances, with the objective of providing an assessment of political stability; security of investment; attitude toward private and (particularly) foreign investors; and adherence to laws, regulations, rules, and agreements.
- Economic conditions, with the objective of assessing likely changes in these conditions and the possible impacts on market demand and price of project output as well as on the prices of project inputs (including labor costs) and the foreign exchange rate.
- Energy sector development patterns, with the objective of forecasting the overall demand-supply balance and associated physical, financial, and institutional constraints on the supply of various forms of energy.
- Subsector issues, with the eventual objective of assessing market demand for project output and issues project sponsors have to face during the construction and operation phases of the project.

The Political System and Circumstances

The political analysis assesses political stability and the degree of government intervention in the economy. In terms of the overall political matters, it is important to provide concise information about

- The relative powers of the legislature versus the executive branch.
- The center or centers of power in the executive branch.
- Turnover in the executive branch.
- The significance of opposition parties and grassroots participation.
- The government's manner of handling the opposition.
- The country's attitude and participation in regional and international cooperation.

With regard to government intervention, the analysis should explain the relevant laws, regulations, and track records in

- Issuing licenses for major investments to private and foreign investors.
- Intervening and exercising control when private and foreign investors are involved.
- Restricting payments of dividends to private shareholders.
- Procuring goods and services when the government is involved in financing, guaranteeing, or otherwise sponsoring foreign borrowing.

Economic Conditions

Macroeconomic conditions can be analyzed to assess the strength of domestic markets, the levels of costs and prices, the country's foreign debt position, and the availability of funds on domestic capital markets.

In order to assess the economic factors affecting market demand, an analysis of historical and projected trends should be carried out for GNP, national consumption, GDP growth rate, and sectoral composition of GDP.

The economic indicators related to costs and wages include inflation indexes (GDP deflator, consumer price index), wage indexes, and deflators for fixed capital formation and investment expenditures. It is also necessary to assess the government policy and track record in combating inflationary pressures.

The foreign debt position of a country is an important determinant of the country's official and unofficial credit rating. The amount and evolution of foreign debt should be explained by examining import and export trends, capital flight, and the resultant accumulated foreign debt. It is also important to determine the portion of the debt that is in the form of public and publicly guaranteed debt and the portion in the form of private nonguaranteed debt.

Finally, discussion of economic conditions should briefly cover interest rates, the banking system, the significance of the domestic capital markets, and any precedents for mobilizing funds for major projects through public offerings or private placements of equity shares or bonds.

The Energy Sector

Analysis of the energy sector should develop a framework for assessing competition among various forms of energy and the advantages, constraints, and resource requirements for various energy supply options. The analysis should cover the following areas:

- Significance of the energy sector to the economy. This significance is assessed both in terms of the ratio of the energy sector's output to total GDP and in terms of the percentages of national investment and public budget used in the sector.
- Energy resources, including recoverable amounts of oil, gas, and coal, as well as hydro potential.

- Sector organization and management. The institutional structure should be discussed with the objectives of indicating the relative shares of public and private sector participation in the supply of energy and describing how the government interacts with the energy sector. The discussion should include the roles of various government agencies in the energy sector.
- Energy demand, including an analysis of historical patterns and forecasts for 15 to 20 years. The analysis should be based on the sectoral (residential, commercial, and industrial) composition of energy demand and should be broken down by type of primary energy (oil, gas, hydro, coal, nuclear) and final energy (petroleum, gas, and electricity).
- Management of energy demand, including regulation of energy prices and promotion of energy conservation.
- Current capacity and sources of supply, including installed capacity for production of oil and gas and generation of power as well as sources of imports of crude oil or petroleum products.
- Constraints and issues in expanding energy supply, including possible insufficiency of domestic energy resources, heavy burden of energy investments on public budget, subsidies and cross-subsidies in energy prices, and corresponding distortions and environmental constraints.

The Relevant Subsectors

The subsector analysis should include

- The significance of the subsector to the economy. Each of the subsectors—petroleum, gas, and power—serves a special role in the economy. The petroleum sector is normally important because it yields royalty and tax revenues. The power sector is important because it facilitates economic growth and development. The gas sector has some of both features and may provide environmental benefits.
- The structure of the subsector. For each of the subsectors, it is important to assess the extent and potential for competition. The current sector structure—for example, the extent of integration, the number of companies, and the share of the largest company or companies—is an important indication of potential competition.
- The government role. It is said that governments do three things: tax, spend, and regulate. Within each of the energy subsectors, these functions are normally determined in the legislative and regulatory framework. The legislative framework determines, through specific laws (for example, petroleum act, gas act) provisions regarding ownership and operation of energy sector resources and facilities. The regulatory framework includes procedures for implementing legislative provisions. If the legislative and

regulatory frameworks are well established, the important parameters of a proposed project can be determined. However, most developing countries have no comprehensive legal and regulatory framework in place. Therefore, many key relationships have to be worked out through specific agreements. Nevertheless, an assessment should be made of applicable taxes and achievable prices and their foreseeable evolution.

- Relevant experience of others. Experiences in the subsector (for example, projects undertaken by oil companies or private power producers) are analyzed with the objectives of identifying behaviors characteristic of past success and outlining lessons that could facilitate preparation and implementation of the proposed project.
- Analysis of market demand and supply capacity. This is obviously the core of the subsector analysis and should cover (1) a historical perspective of demand and supply, (2) the forecast of demand over 15 to 20 years under various assumptions about economic growth and energy prices, (3) planned investments, and (4) the projected supply deficit.

THE INTERNATIONAL BUSINESS ENVIRONMENT

The international business environment can have varying degrees of relevance depending on the type and, particularly, the sources of project inputs and markets for project output. The main areas of analysis include

- World oil markets; international energy prices; and, if applicable, markets for exporting project output.
- Sources of capital equipment; interest rates; and availability of funds from multilateral, bilateral, and commercial sources, including suppliers' credit.

World Oil Markets and International Energy Prices

The depth of analysis of international energy markets depends on whether the project output is fully or partially aimed at export markets. If export is not relevant, then objectives of the analysis would be

- To provide an outlook of international energy markets and the evolution of energy prices. These prices are normally necessary for project analysis because, even if project output is not exported, the domestic price would have some correlation with the international price. In addition, in certain projects (for example, power generation) the fuel may have to be procured from international markets or bought domestically at prices related to international prices.

Figure 8.3
Analysis of Country's Business Environment

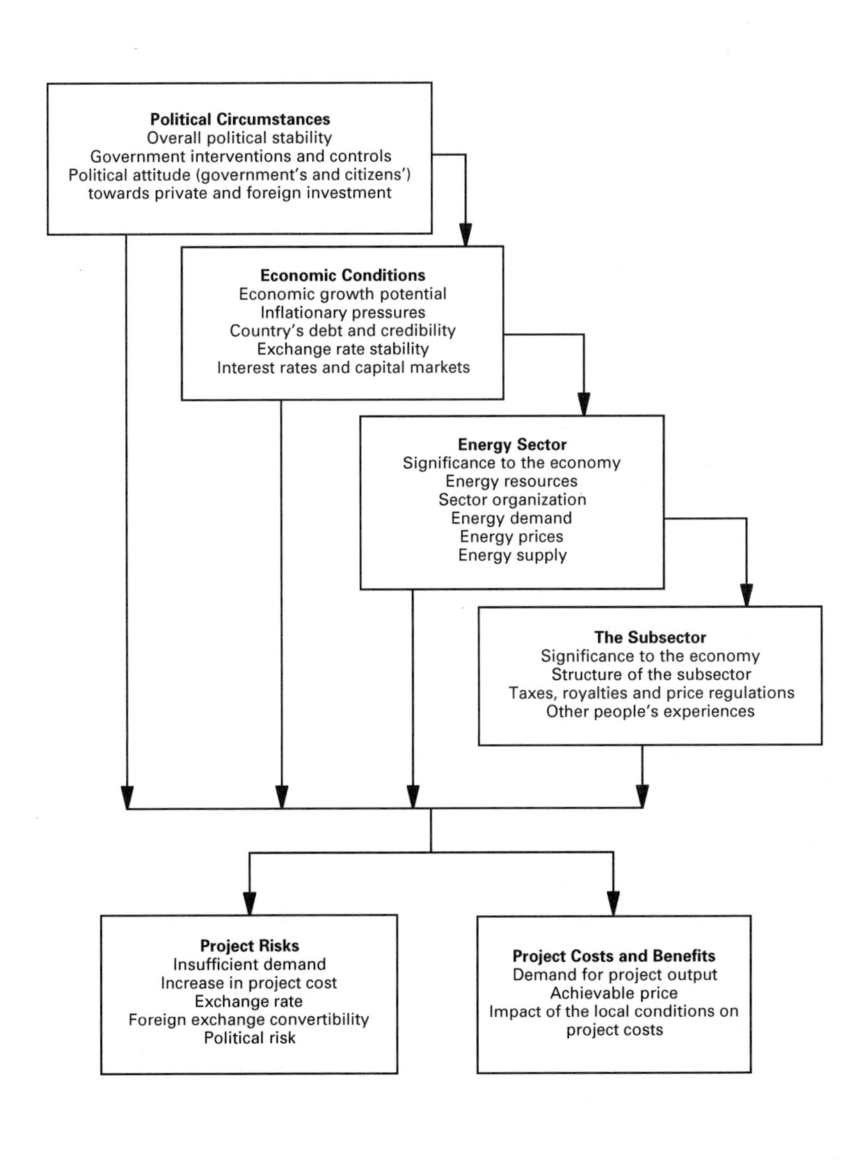

- To assess the correlation between domestic and international energy prices. Domestic energy prices have been subsidized in many developing countries. The extent of subsidy varies greatly among countries, but it has been generally reduced during the second half of the 1980s and the first half of the 1990s. Moreover, most countries plan to remove overall price subsidies, although some cross-subsidies will remain. A further issue in investigating the relationship between domestic and international energy prices is the status of price deregulation.

In the event that output of the proposed project is destined for export, the analysis of international energy markets will have to be much deeper, covering

- Identification and evaluation of specific geographical markets targeted for sale of project output.
- Expansion plans and comparative advantages of suppliers active in targeted markets.
- Legal and regulatory issues relevant to imports.
- Justification that project output can be successfully marketed at the expected price level. Often, there is a need to secure a long-term take-or-pay contract with credible purchasers. The arrangement could also be acceptable if the take-or-pay contract is made with a reputable trading house.

Sources of Purchasing Capital Equipment

The energy equipment supply market is quite competitive. The competition is not limited to price and quality of service. Equipment suppliers often take a flexible approach and do whatever it takes to win business. They could become partners in the proposed energy project, finance purchases, or influence others who finance the project.

Analysis of the equipment supply market covers

- Identification of suppliers with compatible technology and track records in providing satisfactory service during plant construction and operation.
- Examination of the corresponding countries' bilateral facilities including aid, soft loans, and guarantee programs. It is also important to review the decision-making processes involved in these programs. Of particular interest would be the influence each supplier may have in such decisions.
- The financial capacity of equipment suppliers and their normal practice in providing financial assistance in the form of partnership, loan, or credit.

Sources and Costs of Finance

A logical extension of the analysis of the previous section would be review of the capital markets. This review should include

- The availability of funds suitable for the proposed project. This analysis would draw on experience of financing similar projects and assess the preparedness of multilateral, bilateral, and commercial financiers to support the proposed project.
- The cost of capital—for example, evolution of interest rates for each source or method of financing.

THE USE OF GUARANTEE INSTRUMENTS

Analysis of the business environment reveals a host of risks that could endanger a proposed project. Before analyzing project costs and benefits, sponsors should utilize available guarantee or insurance schemes to reduce these risks.

Many methods and a variety of sources are available to guarantee virtually all significant transactions. In the absence of such guarantees, project sponsors themselves must become guarantors of all risks. This creates a number of problems. First, the risks may simply be excessive for sponsors who are primarily in the business of building projects rather than taking risk. Risk taking is now a business of its own, which operates within a very sophisticated array of parameters. Second, even if sponsors are willing to assume all the risks, lenders may not feel sufficiently secure to finance the project. Project sponsors have three main incentives for obtaining third-party guarantees:

- Shifting some of the project risks to other parties.
- Reducing the impact of debts of the project company on their own balance sheets and credit ratings.
- Persuading lenders to extend financing to the project.

Some minor guarantees are available from commercial guarantors such as banks (non-United States), insurance companies, and investment companies. These are normally in the form of letters of credit, which are provided based on the credibility of project sponsors. However, the main potential guarantors are parties with interests in some aspect of the project. These include

- Governments and development agencies interested in the economic impact of the project. They may be persuaded to provide guarantees

Figure 8.4
The Impact of the Business Environment on Project Viability

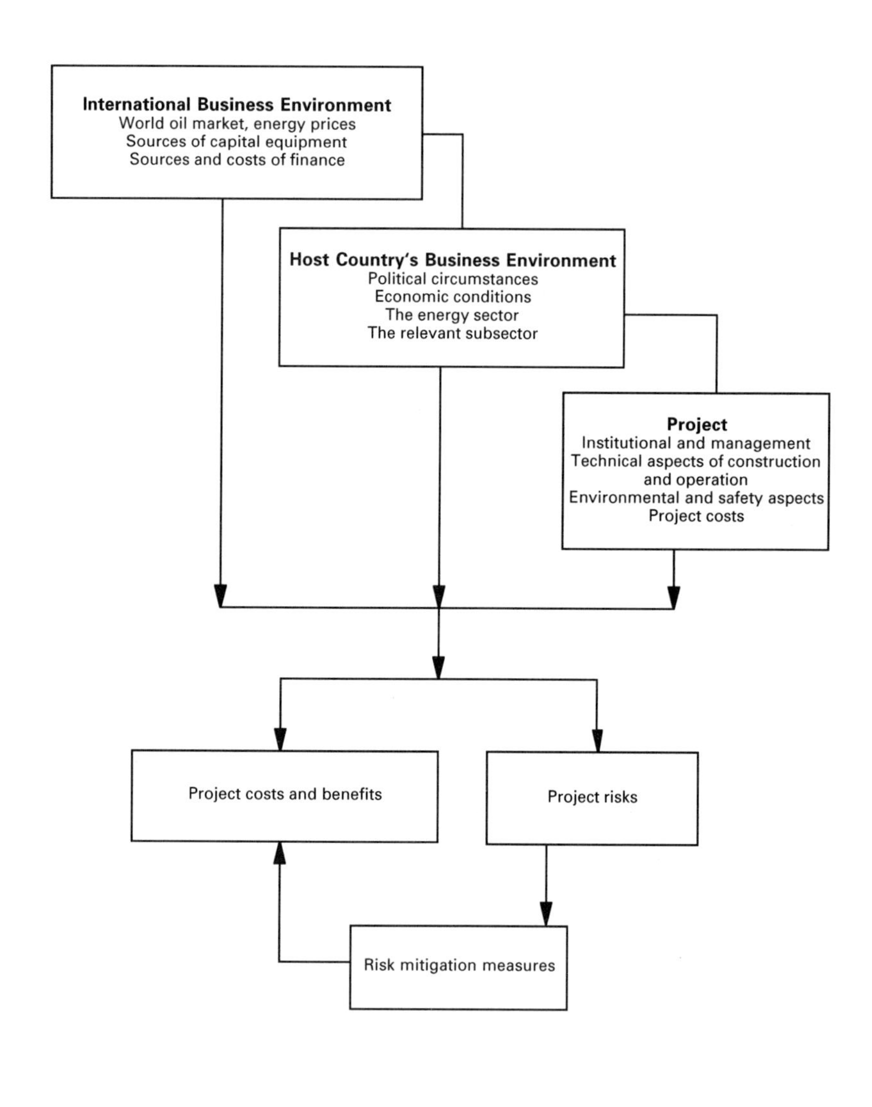

against political risk, problems of foreign exchange convertibility, and aspects of market risk by, for example, underwriting take-or-pay contracts. In addition, a large number of agencies sponsored by the governments of industrialized countries provide bilateral insurance against political risk of investment and finance committed by their own nationals in developing countries.

- Suppliers of equipment and services, including contractors, who want to sell to the project company during the construction phase. They will normally guarantee the cost, timeliness, and performance of their equipment or service.
- Suppliers of raw materials and fuels, who want to sell to the project company during its operational phase. They are normally willing to guarantee the delivery and the price (or a price formula).
- Consumers of the output of the proposed project. They may be willing to sign a take-or-pay contract for purchasing the output. This type of arrangement is viewed as an indirect guarantee and is taken into consideration by lenders and credit rating services.

Guarantee arrangements relating to commercial risks (that is, project completion, cost overrun, delay, fuel supply, operation and maintenance costs, and market demand) are normally arranged during project preparation between project sponsors and other participants. However, guarantees against political risks (that is, expropriation, nationalization, confiscation, currency inconvertibility, labor behavior, and government regulations) are much more delicate and difficult to handle.

Mitigating risks requires considering all possible sources of guarantee and choosing a mix of instruments to enhance prospects for project financing at the lowest possible cost to project sponsors.

These guarantee instruments are classified in several different ways. The following paragraphs describe guarantees as applied, respectively, to commercial aspects of project construction and project operation and to mitigating political risk.

Table 8.2
Types of Project Risks and Parties Influencing Risk

Type of Risk	Parties Influencing Risk
Commercial Risk	
CONSTRUCTION RISK	
Cost overrun	Contractor
Completion delay	Contractor
Increased financial costs	Banks, other lenders
OPERATION RISK	
Unsatisfactory plant performance	Contractor, operator
Excessive maintenance costs	Operator
Fuel supply or fuel cost for power plants	Fuel supplier
Insufficient reserves for oil and gas projects	Sponsors
REVENUE RISK	
Insufficient sale volume	
Low prices	Purchasers of project output
Political Risk	
REGULATORY RISK	
Changes in regulatory regime, including price, environmental obligations, and so on	Government
TRANSFER-OF-PROFITS RISK	
Foreign exchange inconvertibility, restrictions in transferring funds	Government
OTHER	
Expropriation, war, civil unrest	—
Force Majure Risk	
Natural disasters	—

The guarantee instruments for dealing with construction risks fall into the following categories:

- **Sponsors' completion guarantee.** An important period of risk in a project financing is during the construction and startup phases of the project. Financiers would like to receive various types of assurance that these risks are managed. The first and most important assurance is that project sponsors themselves provide a project completion guarantee, which commits the sponsors to completing the project within a certain time period and to providing funds to pay all cost overruns. However, other guarantees, by contractors, equipment suppliers, and so on (as described below), are normally sought to protect the project sponsors and to provide financiers with additional assurances that project completion risks are managed.
- **Lump-sum turnkey contract.** Lump-sum turnkey (LSTK) engineering, procurement, and construction (EPC) contracts are used to ensure that the project is completed on time, within budget, and with acceptable standards of operation. These contracts have stringent reward and penalty clauses that protect sponsors against cost overruns and operational deficiencies.
- **Bid bond.** At the outset of a project, contractors are invited to bid for the project. A bid bond is required of bidders on a contract to ensure that each bidder is serious and would accept the award of the contract if offered.
- **Performance bond.** A performance bond provides additional funds in the event the contractor fails to perform for any reason. The existence of such a bond is also an endorsement of the credit and confidence of the guarantor in the ability and professional standing of the contractor.
- **Advance payment guarantee.** An advance payment guarantee assists the contractor in purchasing and assembling the materials, equipment, and personnel necessary to begin construction, so as to meet the requirements for receipt of progress payments under the contract.
- **Retention money bond.** It is common practice for the sponsors for whom a project is being built to retain or hold back a portion of the progress payment that would otherwise be due, in order to provide a fund to cover unforeseen expenses caused by any contractor mistakes in construction. Because most contractors prefer to receive the progress payments as quickly as possible, they substitute a retention bond for the amount of the funds retained, to receive immediate payment.
- **Maintenance bond.** The purpose of the maintenance bond is to provide a source of funds for correcting construction or performance

defects discovered after completion of construction. Typically, the performance bond and the retention bond are converted to maintenance bonds on completion of the contract.

- **Financial derivatives.** Financial derivatives, such as swaps and options, can be used to hedge against unprotected increases in financial costs, particularly those caused by an increase in interest rates or an adverse movement in the exchange rate. Both of these factors can significantly affect the cost of the project, particularly if sponsors have borrowed funds at floating rates and in a single currency. Interest rate swaps and caps as well as currency swaps and options may be used to deal with these risks.

In addition to well-designed EPC contract protection against operating deficiencies, guarantee instruments for dealing with commercial aspects of operational risks include the following:

- **Put-or-pay contract.** Any foreseen change in availability and cost of energy, raw materials, or product of a project will affect the project's performance and profitability. Most projects hedge against these risks by utilizing put-or-pay contracts. Put-or-pay contracts or supply-or-pay contracts are provided by suppliers of energy, raw materials, or products to projects needing assured supplies of such inputs over long periods, at predictable prices, to meet production cost targets. The put-or-pay obligor must either supply the energy, raw material, or product, or pay the project company the difference in costs incurred in obtaining the input from another source.
- **Take-or-pay contract.** A take-or-pay contract is an unconditional contractual arrangement between the project sponsor and project's customer. The contract obliges the customer to make periodic payments in the future in return for fixed amounts or quantities of products, goods, or services at specified prices. The obligation to pay is unconditional and must be fulfilled whether or not the service or the product is used by the customer. As the cost of service may rise over time because of inflation, the payments are usually subject to escalation. The take-or-pay obligor can protect its interest by retaining rights to take over the project in the event of failure by the supplier to perform. Any such takeover would be subject to the take-or-pay obligor assuming or paying the debt used to finance the project. Examples of take-or-pay contracts can be found in almost all segments of the energy market.
- **Throughput contract.** The equivalent to take-or-pay contracts for projects that provide services such as power transmission, power distribution,

oil pipeline transportation, or refining is called a throughput contract. Lenders regard throughput contracts as a guaranteed source of income; the guarantee is unconditional and lasts for the life of the loan. The obligor pays whether the service is used or not. This type of obligation is sometimes called a tolling agreement, a cost of service tariff, or a deficiency agreement.

- **Escrow account.** An escrow account is not a guarantee but a means of payment, a way to put aside the money needed to repay a debt. The escrow account is a special account, often outside the host country, to which a certain portion of project revenues are channeled. The account is managed by an agent in accordance with a specific agreement. If all project revenues are deposited in the account, then the escrow agent disburses the funds with a structured priority, first covering the operating cost and then covering debt service, before remitting the remainder to the project company. The escrow account agreement terminates when all debt obligations have been fully paid.

The guarantee instruments for dealing with political (country) risks are more complex because of difficulty in precise definition and assessment of resultant defaults. In a broad sense political risks include

- Currency risks—for example, inconvertibility, devaluation, and restrictions on currency imports and exports.
- Taxes and duties—for example, increased taxes on property, production, income, and profits or increased import or export duties.
- Labor risks—for example, changes in laws and regulations dealing with work permits for imported labor, labor unions, and labor compensation.
- Government-intervention risks—for example, local government and federal government interference or harassment through licenses, regulations, police, and military.
- Losses from expropriation, nationalization, confiscation, war, or revolution.

No guarantee instrument covers all of the above risks. Often, project sponsors combine several guarantees to ensure sufficient protection against political risks. For example:

- Guarantees by the host government are often needed to assure the project company that the government will take measures to protect or enhance the interest of the project, provided it is within the government's control and the company functions within the country's legal

Figure 8.5
Risk Mitigation and Guarantee Instruments

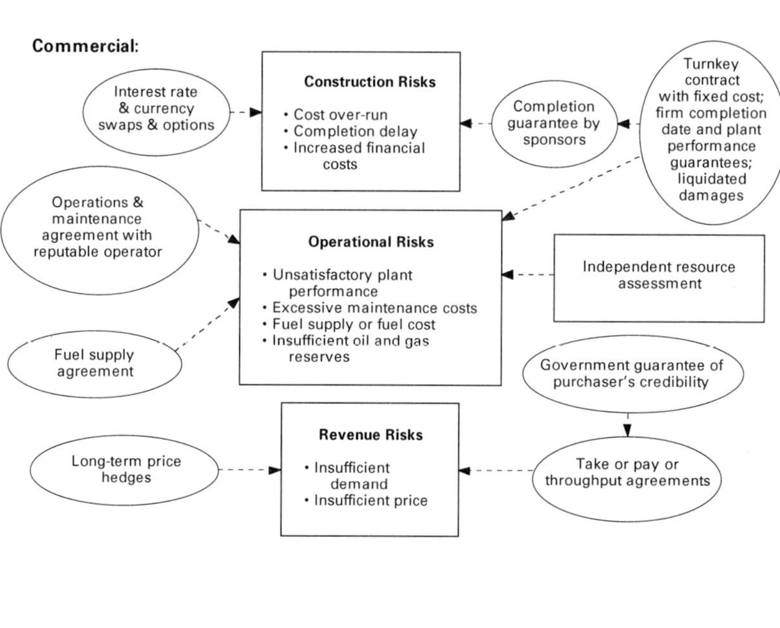

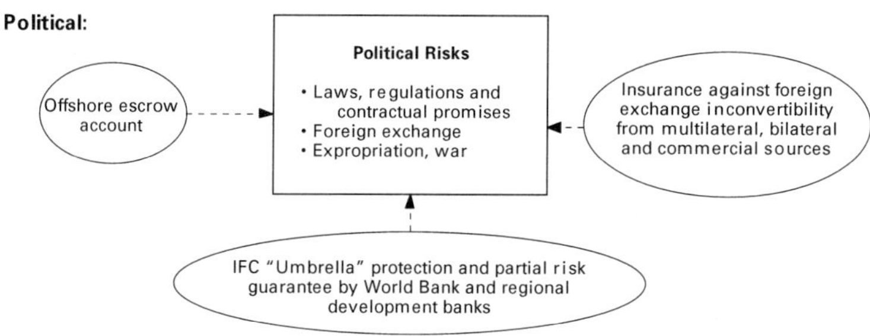

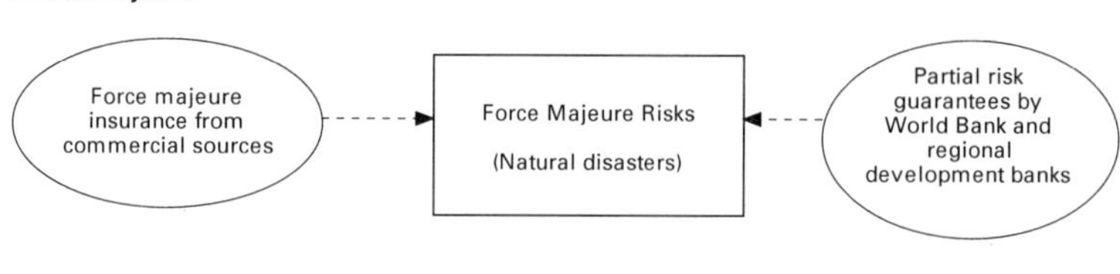

framework. Host-government guarantees against political risk can take many forms, including concession agreements covering some of the risks outlined above, coverage against expropriation or nationalization, and guarantee by the central bank of the host country against risks such as currency and foreign exchange restrictions.

- The Multilateral Investment Guarantee Agency, an affiliate of the World Bank, provides investment insurance against political risk in developing countries.
- The International Finance Corporation, another affiliate of the World Bank, provides various types of comfort to other investors and financiers through its "umbrella" arrangement.
- The World Bank itself may provide partial guarantee against the risk arising from nonperformance of sovereign contractual obligations and from the force majeure aspects of the project.
- Regional development banks (such as Inter-American Development Bank, European Bank for Reconstruction and Development, and Asian Development Bank) provide various types of guarantees against political risk.
- Most export-import banks of industrialized countries provide various types of guarantees against commercial and political risks.
- Most industrialized countries now have specialized government-sponsored agencies that guarantee investments and financing of their own nationals in developing countries.

Table 8.3
Major Agencies Providing Political Risk Guarantee

Multilateral Investment Guarantee Agency (MIGA)	
Political Risks Covered	Insures loans and equity investments against currency convertibility and transfer risk,the deterioration in investors' legal ability to convert and take out profits, debt service and other remittances from local currency. Also insures against loss of assets resulting from full or partial expropriation, nationalization or confiscation by a foreign government; creeping expropriation, such as a series of discriminatory taxes or laws by a government entity over time; and loss of assets or income caused by war, revolution, insurrection or politically motivated civil strife, terrorism or sabotage.

Multilateral Investment Guarantee Agency (MIGA) *(continued)*

Eligibility	Projects in member developing countries are eligible for coverage of new equity investments and related loans for green-field expansion or privatization projects that are economically and environmentally sound.
Restrictions	Will cover up to 90 percent of the risk for both equity and loans.
Maximum coverage	$50 million per project.
Maximum term	20 years

World Bank

Political Risks Covered	Guarantees commercial loans to the private sector against currency convertibility and transfer risk and against breach of contract by the host government or its agencies. Will also compensate for losses caused by a change in government regulations if the change will affect repayment of debt. No coverage of loss resulting from war, political violence or expropriation.
Eligibility	Loans to private companies in its member developing countries.
Restrictions	Requires counterguarantee from host government. Project must be economically viable and comply with the Bank's country assistance strategies and environment requirements.
Maximum coverage	No limit. Prefers to be lender of last resort.
Maximum term	No limit

Asian Development Bank (ADB)

Political Risks Covered	Nonfulfillment of specified government obligations, such as maintenance of an agreed upon regulatory framework, delivery of inputs or payment for outputs by state-owned entities, provision of specified infrastructure for a project, compensation for delays caused by governmental actions or political force majeure and currency convertibility.
Eligibility	Any project eligible for an ADB loan.
Restrictions	The local government will sometimes be required to give a counterguarantee.

Asian Development Bank (ADB)
(continued)

Maximum coverage	No limit
Maximum term	No limit

Inter-American Development Bank (IDB)

Political Risks Covered	Guarantees loans to private sector firms in emerging markets against currency convertibility and transfer risk and against abrogation of government contract obligations, such as a project purchase agreement of concession contract. Compensates for loss caused by change in government regulations if it will affect repayment of debt. Does not cover expropriation or loss resulting from war or political violence.
Eligibility	Projects in the 26 member countries of the IDB. Initial emphasis is to provide guarantees for loans to finance infrastructure projects.
Restrictions	Requires counterguarantee from the host government.
Maximum coverage	No absolute limit, but can't cover all the debt of a project.
Maximum term	15 to 20 years

Overseas Private Investment Corporation (OPIC)

Political Risks Covered	Insures loans and equity investments against currency convertibility and transfer risk, the deterioration in investors' legal ability to convert and take out profits, debt service and other remittances from local currency. Also insures against loss of assets resulting from expropriation or creeping expropriation, nationalization or confiscation by a foreign government and loss of assets or income caused by war, revolution, insurrection or politically motivated civil strife, terrorism or sabotage.
Eligibility	Equity investments in and loans to projects owned by U.S. citizens and U.S. companies and their foreign subsidiaries.
Restrictions	Projects must be in one of the 145 countries that have signed bilateral protocols with the U.S.

Overseas Private Investment Corporation (OPIC) *(continued)*	
Restrictions *(continued)*	They must show potential to provide social and economic benefits to the host country and may not harm the U.S. economy.
Maximum coverage	Insures up to $200 million per project or up to 90 percent of equity investments and 100 percent of loans, whichever is less. In projects with many investors, may insure no more than the percentage of debt or equity owned by U.S. investors.
Maximum term	20 years
Export-Import Bank of the U.S. (US Exim)	
Political Risks Covered	Same as OPIC.
Eligibility	Covers U.S. exports to almost any country.
Restrictions	A small number of countries, among them Cuba are legally barred. Provides only political risk guarantees during the construction phase of a project; afterward both political risk and comprehensive coverage are available.
Maximum coverage	None
Maximum term	Consistent with OECD guidelines: 12 years for power projects, otherwise 10 years.
Export-Import Insurance Division of Japan's Ministry of International Trade and Industry	
Political Risks Covered	Currency transfer controls; war, revolution or civil disturbance, other occurrences in host country for which neither the lender nor the foreign party is responsible.
Eligibility	Japanese companies or non-Japanese subsidiaries registered in Japan. One of the few export credit agencies that can cover content originating outside its home country.
Restrictions	No guarantee for loans used primarily to buy Japanese exports or where there is an equity relationship between the lender and the borrower.

Export-Import Insurance Division of Japan's Ministry of International Trade and Industry *(continued)*	
Maximum coverage	No absolute limit, but up to 97.5 percent of loan or investment, according to insurance contract type and country risk.
Maximum term	For untied-loan and investment insurance, 15 years; export-proceeds insurance, 10 to 12 years; other types, up to 5 years. Term may be raised at the request of the minister.
Export-Import Bank of Japan (JExim)	
Political Risks Covered	Only currency transfer and convertibility risk.
Eligibility	Loans from financial institutions in Japan, including branches of foreign banks, to recently privatized businesses and regulated utilities in developing countries; also equity joint ventures between governments and private enterprise.
Restrictions	Not restricted to Japanese export finance. The only export credit agency not restricted to home country export financing.
Maximum coverage	95 percent for a full-length guarantee, 100 percent after an initial uncovered period of $2^1/_2$ years.
Maximum term	12 years
Export Credits Guarantee Department of the U.K. Board of Trade (ECGD)	
Political Risks Covered	Damage and operational losses caused by war or political violence; losses from expropriation, including discriminatory action; currency transfer restriction. Individualized coverage for other specified risks.
Eligibility	U.K. company investments in developing countries.
Restrictions	Only for new direct investments in overseas enterprises or new capital injected into an existing business. No coverage for investments intended to last less than 3 years.
Maximum coverage	£100 million ($155 million) of loan principal or equity investment plus retained earnings.
Maximum term	15 years at set rate with annual option to renew.

Cie. Française d'Assurances Commerciale Extérieure

Political Risks Covered	War, political violence, expropriation and currency controls; nonpayment by a public buyer, such as government and government-owned banks or enterprises.
Eligibility	French companies and their overseas subsidiaries investing abroad. Covers equity and long-term loans. A December 1994 agreement with MIGA will facilitate co-insurance and ease coverage of investments by consortia of French and non-French companies.
Restrictions	Only for countries with which France has bilateral trade agreements, with case be case exceptions.
Maximum coverage	None, but covers only 95 percent of investment.
Maximum term	15 years, with possibility of 5-year extension.

Export Development Corporation of Canada (CDC)

Political Risks Covered	Damage to assets and loss of income from war, revolution or insurrection; loss resulting from government restriction on currency convertibility and transfer and from both overt nationalization and creeping expropriation.
Eligibility	Economically and technically viable projects that meet the environmental and regulatory standards of the host country. Any host country is eligible, subject to judgments on political risk.
Restrictions	Product or service exported to project must have at least 50 percent Canadian content, unless there are indirect Canadian benefits, such as technology transfer, and enhancement of future business prospects.
Maximum coverage	$100 million per project.
Maximum term	15 years

9

ESTABLISHING ECONOMIC AND FINANCIAL VIABILITY

Economic and financial viability are the priority concerns of investors and financiers examining a project. These factors must therefore receive priority attention in the project package. Indeed, analysis of financial and economic costs and benefits is the core of a project document because it brings the results of all other analysis into an all-encompassing framework and indicates the risk-reward profile and therefore the attractiveness of a project.

Computation of economic and financial ratios is based on streams of capital costs, operation and maintenance (O&M) costs, and project benefits (for example, revenue from selling the output). The economic and financial analysis is concerned with

- Reviewing all the factors that affect capital costs, O&M costs, and project revenues and other benefits; quantifying the input of all these factors; and arriving at one set of figures as representing the most likely streams of costs and benefits.
- Calculating relevant economic and financial ratios.
- Carrying out a comprehensive risk analysis based on possible variations in cost and benefit streams.

ECONOMIC VERSUS FINANCIAL ANALYSIS

Economic analysis usually refers to an assessment of the costs and benefits of a project to the society or the country as a whole. *Financial analysis* of a project refers to assessment of the costs and benefits of a project to the company undertaking the project. For example, if electricity consumers receive a benefit equivalent to 8¢/kilowatt-hour (kWh) from using electricity but pay only 5¢/kWh, the economic analysis would use the 8¢/kWh as an indication of benefits, but the financial analysis would use the 5¢/kWh because this is the actual revenue from selling the electricity.

The distinction between economic and financial analyses and benefits was most pronounced during the 1960s and 1970s, when prices were more controlled and distortions were greater. During this period, substantial theoretical work was done to derive methods of quantifying economic costs and benefits. These methods attempt to adjust the observable cost and benefit figures to arrive at the corresponding "socially optimal" values.

Most observable figures are financial costs and revenues. Examples are the actual price received by a utility for selling electricity, the actual exchange rate at which the company converts foreign currency, and the actual wage paid to the laborer. These figures may not indicate the true costs and benefits to the economy, however. First, taxes, royalties, economic rents, and so on are all considered as cost to the company but are indeed money received by the government; therefore, these components do not represent a true cost to the economy. Second, market distortions caused by controls, subsidies, and other market failures result in prices, exchange rates, and interest rates that do not correspond with competitive market equilibria.

Adjustments of the observable cost and benefit figures reflecting taxes and royalties are rather straightforward and are routinely done in project analysis. The adjustments aimed at compensating for market distortions are more complicated. They normally involve estimating shadow prices, shadow exchange rates, shadow wage rates, and so on, which reflect the economic values of these parameters. This latter type of adjustment became less important in the 1980s and 1990s, as most developing countries moved toward market economic systems, and as energy prices and exchange rates became more market-related.

For many energy projects, an additional adjustment should be made to take account of a "depletion premium" if the proposed plant produces or uses exhaustible resources. Again, in an economically efficient environment, financial prices incorporate this factor.

Economic and financial analysis should start with financial costs and benefits: that is, the company's expenses and revenues. These figures form the basis of financial analysis. In contrast, economic analysis should

- Deduct from project costs funds paid to the government for royalty, production tax, asset tax, import tariffs, and so on. These items, although representing costs to the company, are not a cost to the economy. They are simple transfers of funds from the company to the government.
- Add to project revenues the sales tax collected from consumers. Again, this amount does not represent a benefit to the company but is part of the overall project benefits for the economy as a whole. In other words, the revenue from the project includes all the money paid by consumers, whereas part of this revenue is transferred to the government and the rest remains with the company.

Incremental versus Total Costs and Benefits

The analysis of costs and benefits of a project should be based on the project's corresponding costs and revenues. In certain investments—for example, construction of a new power plant or a grassroots refinery—the project is an independent facility with clear and measurable output. In some other investments—for example, expansion of power or gas distribution systems—the project adds components to an existing facility. The primary attempt is normally aimed at assessing and analyzing "incremental" costs and benefits: that is, those associated with the proposed investment. When this is not possible, the analysis is based on total facilities.

The Methods and Ratios

Two key measures used by investors to assess the economic and financial viability of a project are the *net present value* (NPV) and the *internal rate of return* (IRR) of the project. The NPV of a project is the discounted value of cash inflows less the cash outflows of the project. In order to calculate the project NPV, one has to assign a value to the discount rate. This value, a market interest rate relevant to the project, indicates the cost of money. Thus, NPV measures the actualized net revenue (that is, the revenue less all the costs) and indicates the attractiveness of an investment to the sponsors. In other words, NPV indicates the difference between an investment's market value and its costs and, thus, the reward for sponsors for putting capital at risk to undertake the project. For example, an NPV (at a 10 percent discount rate) of $20 million indicates that not only all the initial project expenditures, including the capital funds put in by sponsors, earn a 10 percent annual return but also that there is an additional net revenue whose present value is $20 million.

The IRR is the discount rate that makes the NPV of a project zero. An NPV of zero indicates that the project will earn a rate of return equal to the discount rate.

Lenders' return on a project (that is, interest received on the loan) is determined in advance and does not change if the project turns out to be more profitable or less successful than initially appraised. Lenders have legal assurances, based on well-designed security packages, that principal and interest on the loan will be paid. Nevertheless, they examine carefully the NPV and IRR to make sure that project sponsors have a strong financial incentive to undertake the project and see it through to completion.

The IRR is further analyzed to derive the appropriate numbers for return on equity investments. The IRR measures return on the project as a whole. If the project is funded through 30 percent equity and 70 percent debt, the project's IRR will approximately follow the following formula:

Project IRR = 0.3 (Return on equity) + 0.7 (Average interest rate on debt)

For example, if the project IRR is 15 percent and the average interest on the debt is 10 percent, then the return on equity is about 27 percent.

The desirable IRR depends very much on project risks and on whether the project is undertaken by private or state sponsors. Private sponsors want at least 25 to 35 percent return on equity. This translates into a project IRR of above 15 percent. State sponsors normally require about 10 percent return on equity, which would translate into a 10 percent project IRR assuming the interest rate is also 10 percent.

State-sponsored investments will also emphasize economic IRR. The general guideline is that the economic IRR of the project should not be less than the country's overall opportunity cost of capital, a discount rate normally assessed by the main economic body of the country (for example, the planning agency or ministry of finance).

Lenders derive comfort from sponsor's equity investment in the project at a level commensurate with risk; for example, power plants would support higher "gearing," or debt versus equity ratio, than cyclical refineries or petrochemical plants. However, in limited- or nonrecourse project financing, the value of the asset is tied to the cash flow that can be generated and made available for debt service—that is, scheduled interest and principal payments. Thus, lenders place more emphasis on projected cash flow analysis than on return on investment (NPV or IRR).

Several ratios are calculated based on the project cash flow. For example, (1) the debt service coverage ratio (DSCR)—that is, the cash flow available for debt service, calculated for each year on a cumulative basis; (2) the loan life coverage ratio, which divides the outstanding debt into the remaining cash flow stream; and (3) the interest coverage ratio the ratio of available cash flow over interest payments. The lenders' covenants set certain floor values for these ratios. The floor value, always above 1.0 (for example, 1.3), varies depending on the type of project and financing structure. The floor value may be revisited at project completion when recourse to project sponsors is due to be reduced substantially or eliminated.

If coverage ratios fall below the specified floor levels, lenders have the right to "call" the loan and exercise their lien over the project, after appropriate notice. However, in practice, such situations are usually avoided through preventative measures. For example, lenders and sponsors normally set up a disbursement account, with a "trustee," which receives all project revenues in escrow and pays out funds according to a defined order of priority. Whenever risks regarding foreign exchange availability or convertibility are present, an off-shore disbursement account is established to pay foreign exchange obligations, including interest and principal payments of foreign loans.

UPSTREAM OIL PROJECTS

Economic and financial analysis of upstream petroleum projects involves a number of special issues, the most important of which are choosing an acceptable price forecast for crude oil; accounting appropriately for the size of oil reserves and their producibility; incorporating into calculations the costs and benefits of any associated gas (or liquefied petroleum gas [LPG]); and considering taxes, royalties, and "profit oil."

Price of Crude Oil

Price of crude oil represents the most important piece of information for an upstream petroleum project. However, pricing of crude oil is also a very controversial matter. The controversy emerges from two sources: (1) the differences among various forecasts of the international price and (2) the way that the international price becomes relevant to the project. The first type of controversy has now become part of the petroleum business,

Figure 9.1 Calculation of Net Present Value and Internal Rate of Return

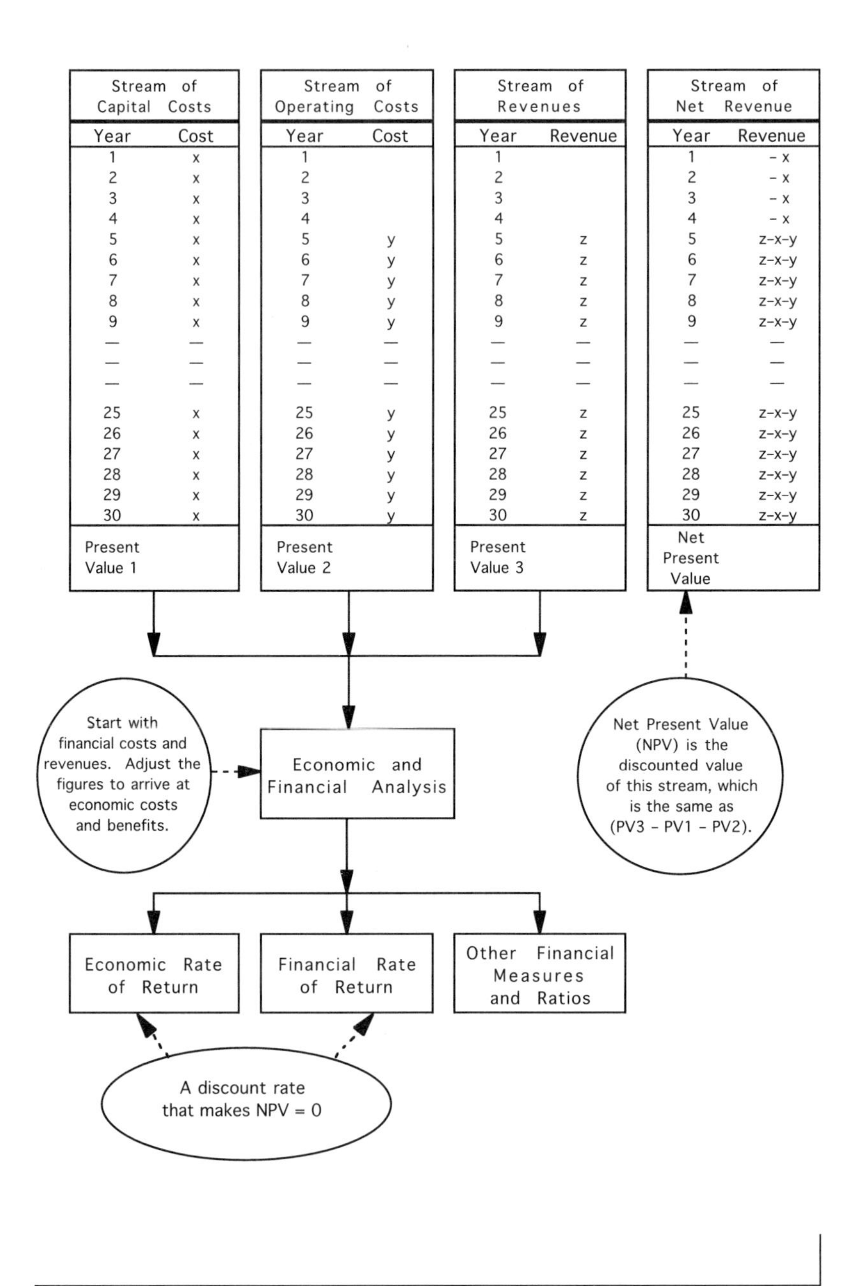

and investors and financiers have learned to live with it. It is an accepted practice to take a reputable published forecast as the base-case scenario and carry out sensitivity analysis to quantify the risks associated with lower prices.

The second type of controversy, the manner in which the international price becomes relevant to the project, must be addressed and resolved before the final application for financing. If all of the project output is exported, the matter is rather straightforward. *Export parity*, the price minus the associated transportation cost, is used as the basis for pricing the crude. However, in most oil projects, the output is either fully or partially sold to domestic refineries. In such a case, planners must estimate the price at which the producer can sell the crude. Financiers would normally want to see that the producer has a firm contract with a clear price, or price formula, with domestic refineries.

The negotiated price between a producer and a refiner depends on the overall demand-supply picture and alternative options for either side. In the event that the proposed upstream project is the only source of domestic crude supply, the producer may force the refiners to pay a price close to import parity; that is, the international price plus cost of transportation. In the event that project output represents one source of crude among many, the refiner likely would not pay more than the export-parity price.

Economic pricing of crude oil, the pricing used in economic rather than financial analysis, is determined in the following manner:

- If the country is short of crude oil and relies on imports, output of the proposed project will replace foreign oil. The benefit to the economy is estimated based on the import-parity price of crude oil.
- If the country has a surplus of crude oil, output of the proposed project would be exported or would replace that of some other producers, who would be forced to export. The benefit of the project to the economy is estimated based on the export-parity price of crude oil.

The Relevance of Netback Value

Wherever the pricing of crude oil is subject to question, the seller and buyer consider the oil's *netback value*. This is an estimate of the value of products that a given quantity of a specific crude will yield after processing by a particular refinery, based on product prices in a particular market. Specifying

Figure 9.2
Petroleum Exploration and Development Projects: Special Issues in Economic and Financial Analysis

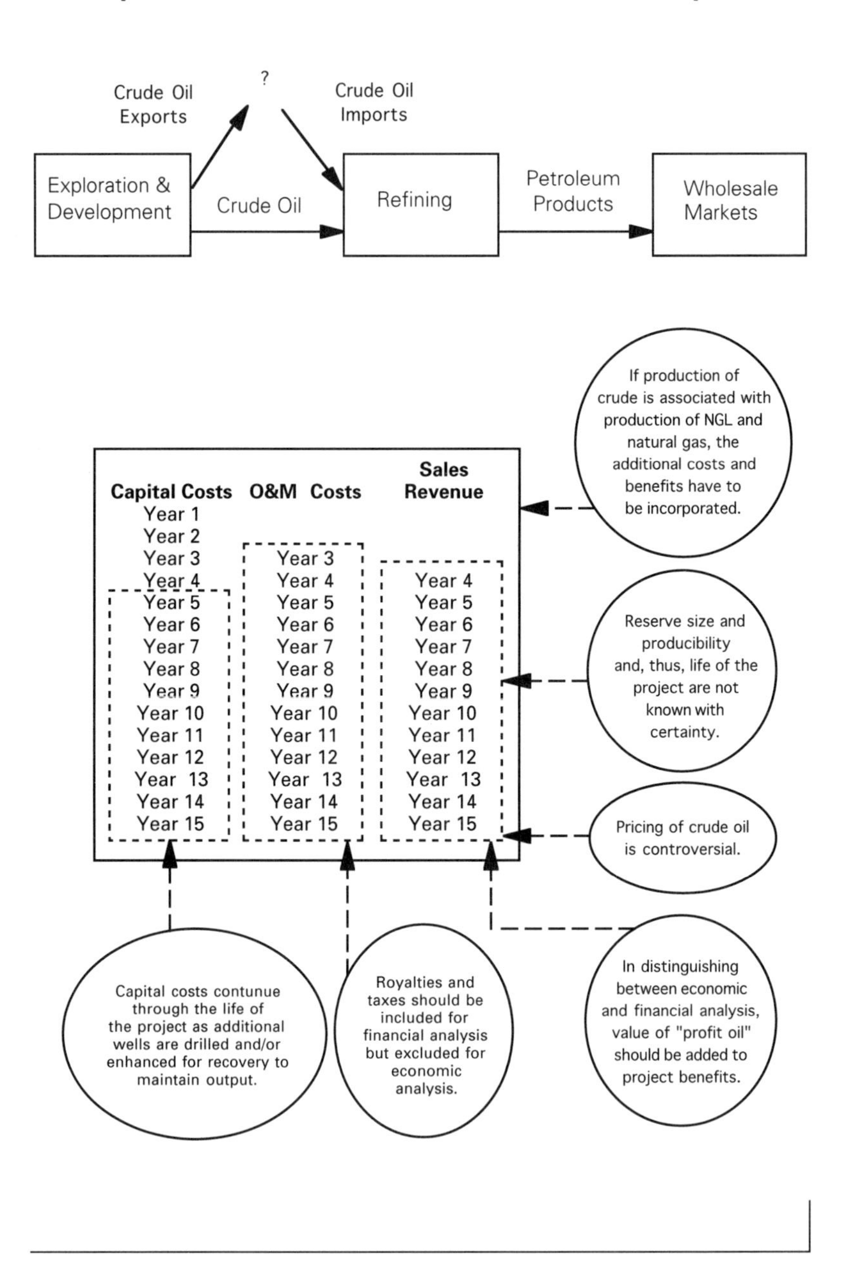

crude type and refinery characteristics is important because variations in either would change product yield patterns central to the calculation.

Steps involved in the calculation of netback value depend on the stages of physical movement of crude oil from the wellhead to the refinery. In a typical case, the physical movement of oil is from field to loading terminal, from loading port to receiving port via tanker, from receiving terminal to refinery, and from refinery to market. Computation of netback value follows these same movements but in reverse order, as follows:

- First, the weighted average value of the refined product obtainable from a barrel (bl) of crude oil at the refinery gate is computed. The result is known as the gross product worth (GPW) of the crude. It is calculated by multiplying the prevailing price for each product by its percentage yield from one barrel of crude oil.
- Then, the cost of refining, which consists of out-of-pocket operating expenses involved in handling the last barrel of crude by a refiner, is deducted from the GPW to arrive at the net product worth (NPW) of the crude. Since the marginal barrel is under consideration, this result does not include any amortization or depreciation.
- Finally, the costs of transportation and insurance are deducted from the NPW to arrive at the netback value of crude at the port of loading.

The most important information in this calculation is the refinery yield—that is, the mix of products obtained from a barrel of crude oil. This mix depends not only on the quality of the crude but also on local demand patterns and the technical capabilities of individual refineries. Thus, looking for specific yield patterns would require compiling thousands of mixes corresponding with various crudes, various refineries, and different periods. Recognizing the impossibility of such a task, the *Petroleum Intelligence Weekly* has developed data on crude oil yield patterns that are typical or representative of the refining industry in each of the six major refining centers.

Second, the refining cost, used in netback value calculations, does not include capital cost. The implication is that netback value is purely a marginal phenomenon that corresponds with the short-term operation of a refinery. Thus, netback value, as calculated and published in industry papers, is not intended as a basis for long-term development and resource management.

Figure 9.3
Netback Value Calculation for Crude Oil

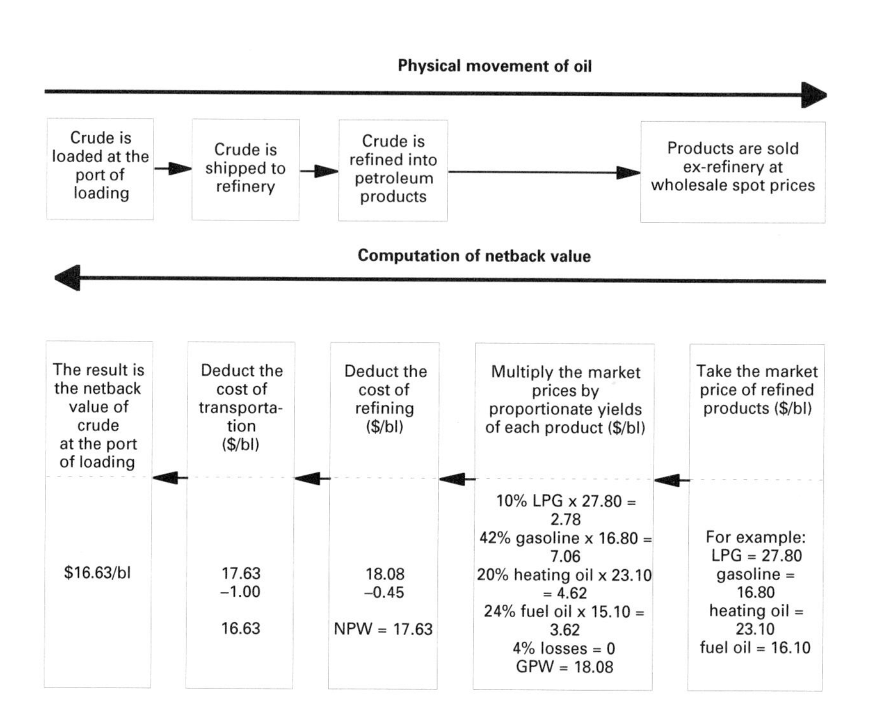

GPW = Gross product worth
NPW = Net product worth

However, a combination of netback value and capital cost of refining may be used in long-term analysis when other measures are not available.

Netback value is often used in negotiations of crude price supplied to captive markets, in which petroleum product prices are not linked to international prices. Refineries in these markets charge product prices substantially different from international prices. Thus, the value of crude oil to these refineries may differ from international crude oil prices. In such situations, netback value of the specific crude oil is calculated based on the refinery yield and prices in the domestic market. The netback value will then become a basis for negotiating the crude oil price.

Size of Oil Reserves

For upstream projects, financing institutions make sure that sufficient oil reserves are present to support projected production profile and, therefore, repayment of the debt. This is usually done by obtaining independent certification of the reserves by an internationally reputable company. Reserves are classified into three categories: (1) proven reserves, which can be recovered with a high degree of certainty; (2) probable reserves, which are likely to be proven but are included in part of the field that requires further delineation or need confirmation of higher recovery by enhanced recovery methods; and (3) possible reserves, which involve high risk regarding recovery. Conservative financiers consider only proven reserves. However, in the early stages of field development it is common to include risk-weighed probable reserves to estimate production profiles and field life. The risk factor applied to these reserves varies between 30 and 50 percent, depending on the degree of confidence in the estimate of the probable reserves.

Other Issues

Economic and financial analysis of upstream oil projects should also consider that

- Capital expenditures normally continue during the life of the project to maintain or increase output, for example, by drilling more wells or implementing enhanced recovery systems and, finally, for decommissioning.
- In certain upstream projects, part of the output is transferred to the government (usually the national oil company) under terms of a production-sharing contract. The value of this amount of oil is not included in project benefits when carrying out financial analysis.

However, the value of this output should be incorporated into project benefits when computing economic rate of return.

- The royalty and the production tax paid to the federal and local governments are included in O&M costs for financial analysis. They should be excluded from O&M costs in calculations of economic rate of return.
- Often an oil field produces crude as well as natural gas liquids (NGL). The NGL output should also be priced appropriately and added to project benefit.
- In certain fields, crude oil is produced in association with natural gas. The "associated" gas is either flared, reinjected, or collected for consumption. In each situation, the costs and benefits have to be incorporated in the analysis (see also the next section, Upstream Gas Projects).

UPSTREAM GAS PROJECTS

Economic and financial analysis of gas production faces issues similar to those discussed in the previous section about upstream oil projects. However, an important parameter for analysis—the price of natural gas—is determined in a different manner from that of crude oil.

Gas Pricing

This is a complex issue because

- Governments often intervene to some degree in determining gas prices. In such cases the issue, even for government authorities themselves, is to find a price level (or price formula) that is fair to consumers and producers and that results in economic production and consumption of gas. Quite often, social and political considerations make the price setting more complex.
- The negotiations between sellers and buyers are based on a host of variables, some of which are subjective and difficult to quantify. This complexity is also highly dependent on the industry structure and the shares of the public and private sectors in the industry ownership.
- No international price, and therefore no generally accepted reference point, exists for natural gas.

A host of factors affect the views of sellers and buyers about the price of gas. For the seller, the price should cover cost of production and (if applicable) of processing and transportation. In addition, the seller will attempt to receive a higher price to cover the intrinsic value (also referred to as economic rent or depletion premium) of natural gas. Further, every seller would want to maximize profits or at least include a just reward for exploration risks. Since a "normal" return is included in the cost of supply, the additional profits are as "above-normal."

For the buyer of gas, the starting point for assessing price is the cost of replaced fuel. If gas is replacing coal, then the price of coal, after certain adjustments to take account of differentials in capital cost and efficiency, would represent the value that the buyer receives from using natural gas. In addition, the user will receive benefits such as environmental advantages, ease of use, and security of supply.

Nevertheless, the buyer does not want to pay for all these benefits. Indeed, the buyer normally argues for a price low enough to provide cost-saving incentives to switch to gas.

Although the above considerations establish a conceptual framework for the discussion of the gas price, calculation of a price figure is not a straightforward matter because (1) several of the above variables (such as economic rent and environmental benefits) are not easily quantifiable and (2) the seller wants to charge a price higher than indicated by the costs, and the buyer wants to pay a price lower than indicated by the benefit of using gas. Thus, in most price negotiations, the seller refers to the benefits the buyer receives from using gas to justify a higher price, whereas the buyer refers to cost of supply to limit the price to out-of-pocket supply costs.

Hence, price negotiations may result in any price between the cost of production and the economic value of gas. The negotiation positions of gas sellers and buyers depend on the structure and organization of the gas sector. In some developing countries, the gas sector is owned and operated by an integrated national gas company. In such a case, the gas price is set by the government (or national gas company), and negotiations take place with large consumers, such as power companies and industrial customers.

In other countries, upstream investments are undertaken by the private sector. The gas price is either announced by the government or negotiated between the producer and the national gas company. In the few countries where transmission and distribution activities are in the private sector, producers negotiate gas prices with distribution companies and sometimes directly with major consumers.

Figure 9.4
Sellers' and Buyers' Views of the Gas Price

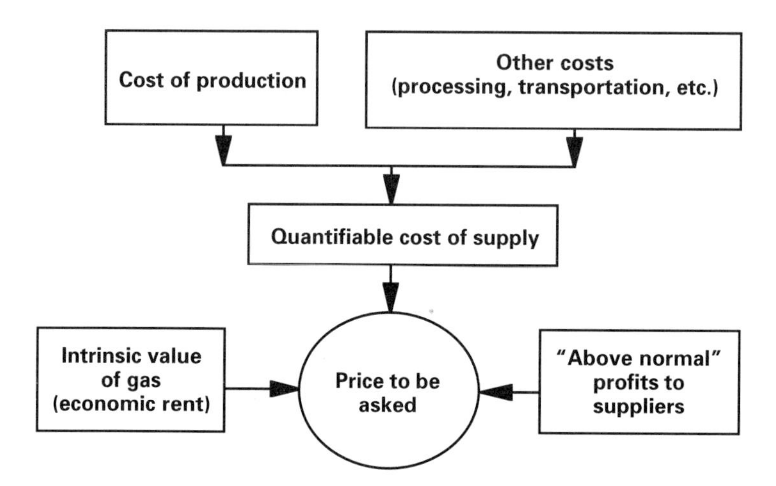

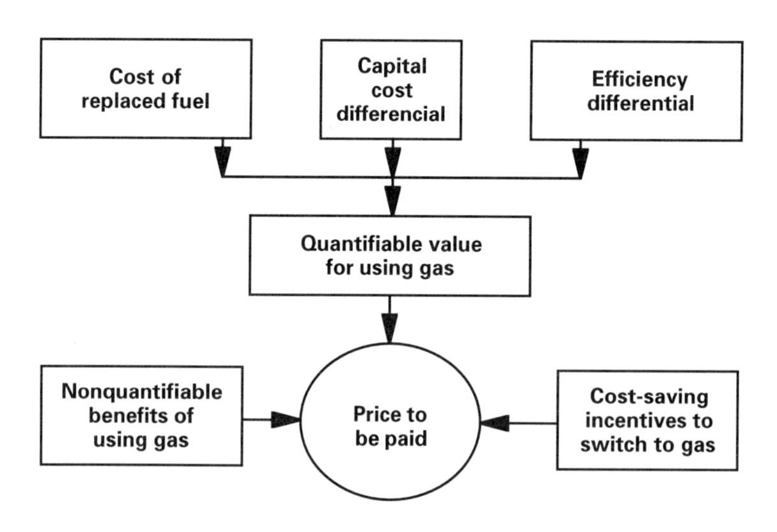

The Relevance of Netback Value

The concept of netback value is often used in price negotiations and by governments in deciding about taxes and other gas-related policies.

For each gas supply system, the netback value can be calculated at various points of delivery. The starting point for calculation of the netback value is the ending point in the delivery system. In other words, the netback value is first calculated for the point of final consumption (at the burner tip) based on capital and fuel costs of the alternative fuel. For example, if gas is replacing coal in power generation, the costs of generating power from a coal-based plant should be compared with those of generating power from a gas-based plant. The netback value for gas is the gas price at which the final costs per kilowatt hour (kWh) would be the same for coal-based and gas-based electricity. Thus, the calculation procedure involves

- Estimating costs per kilowatt hour of fuel and capital for a coal-fired power plant. The sum of these (fuel and capital) costs gives the overall cost per kilowatt hour of electricity generation by the coal power plant.
- Estimating the per-unit capital cost of the gas-based power plant.
- Deducting the second of these from the first; that is, assigning a per-kilowatt-hour fuel cost to gas that would make the overall per-kilowatt-hour cost of gas-based power generation the same as that of coal-based generation.
- Converting the per-kilowatt-hour fuel cost to the equivalent price per million British thermal units (MMBTU, or price per thousand cubic feet [MCF]) for natural gas based on the thermal efficiency of the plant and the heat content of the gas. The resultant figure is the netback value of gas at the point of consumption.
- Netback value can be calculated for other points in the delivery system, such as at the wellhead or ex-transmission, by deducting the appropriate costs. Netback value is normally calculated for alternative uses of gas—for example, gas used in the residential sector, or in the industrial sector (in the form of fuel or petrochemical feedstock). The calculation results in a range of netback values. Theoretically, the economic benefit of additional (marginal) gas supply is indicated by the lowest netback value among the sectors that consume gas. However, in the discussion of gas prices the negotiating parties review the entire range of netback values.

Figure 9.5
Calculation of Netback Value for Natural Gas

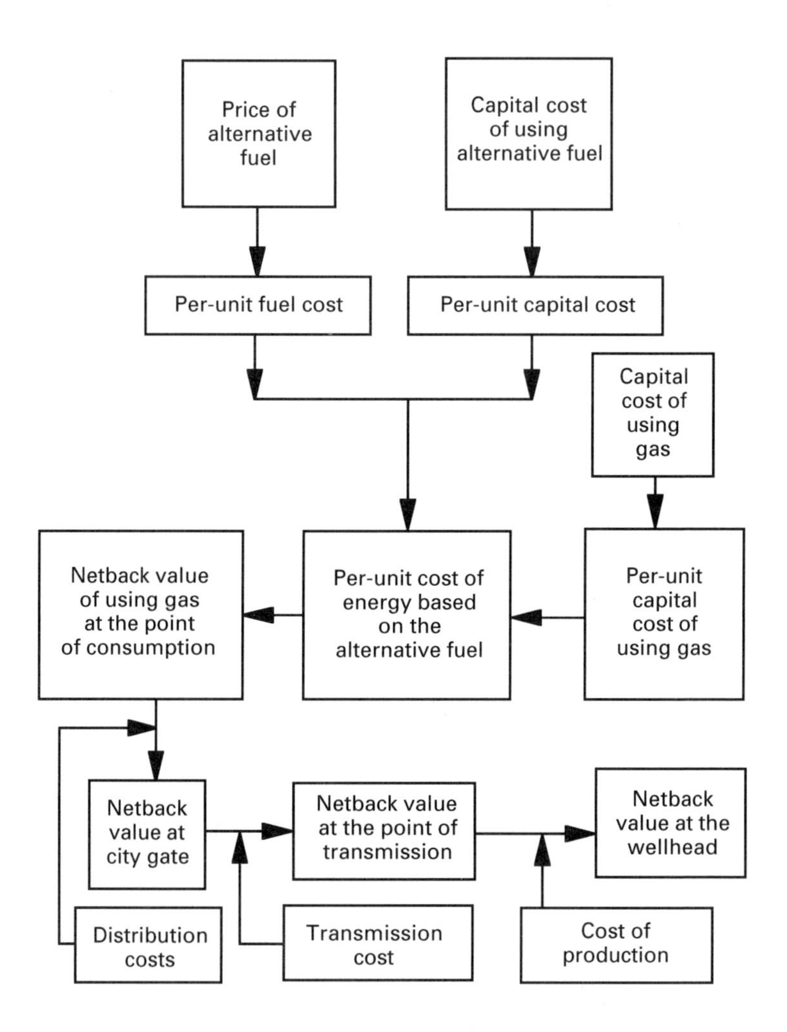

For financial netback value, market prices of alternative fuels and capital equipment are used.

For economic netback value, market prices should be adjusted to exclude taxes and duties.

Table 9.1 Calculation of Netback Value of Gas Use in Power Generation

	Coal-Fired Steam Plant	Gas-Fired Combined-Cycle Plant
Generating Capacity	300 MW	300 MW
Load Factor	75%	75%
Output1971	GWh/y	1971 GWh/y
Thermal Efficiency	34.40%	45.90%
Capital Cost	US$1100/kW	US$650/kW
Plant Cost	US$330 million	US$195 million
Life of the Plant	30 years	25 years
Per Unit Capital Cost	2.2¢/kWh	1.2¢/kWh
Fuel Price	$2.00/MMBTU	$4.60/MMBTU
Per Unit Fuel Price	2.0¢/kWh	3.2¢/kWh
Other O&M Costs	0.5¢/kWh	0.3¢/kWh
Generation Costs	4.7¢/kWh	4.7¢/kWh

Notes: Per-kWh generation of coal-based power is calculated from the data and equals 4.7¢/kWh.
Per-kWh capital and O&M costs of gas-based power are calculated from the plant data and equal 1.2¢ and 0.3¢/kWh, respectively.
The fuel price for gas (3.2¢/kWh or $4.60/MMBTU) is calculated so that the overall cost of gas-based power is the same as that of coal-based electricity.

The Relevance of the Depletion Premium

Like oil, natural gas is often viewed as an exhaustible resource. For this reason, the economic cost of gas supply consists of two components: (1) the cost of gas production (and transmission and distribution if relevant) and (2) a depletion premium indicating the opportunity cost of consuming a unit of the depletable resource now rather than in the future.

Assessment of the depletion premium for natural gas involves the following steps:

- Projecting gas demand and determining the year (switching time) in which further growth in gas demand would not be sustained by the proven gas reserves. The limited availability of gas would then force the country to satisfy incremental demand with an alternative fuel.

- Projecting the price of the alternative fuel to the switching time and adjusting this price to take account of differentials in capital and operating costs and thermal efficiency.
- Calculating the difference between the above adjusted price and the cost of gas production at the switching time. This difference represents the depletion premium at the switching time.
- Computing the depletion premium for prior years by appropriately discounting the premium at the switching time.

As indicated by the above methodology, the depletion premium would be small in gas-rich (gas-surplus) countries. For a gas-short country, the switching point is the present, which means that the calculation of the depletion premium coincides with the calculation of the netback value. Therefore, the concept of the depletion premium provides a general framework that can be used to draw the following conclusions:

- In countries with abundant gas reserves, the depletion premium is negligible. The cost of gas supply is approximately the cost of production. The gas price is set either by the government based on cost considerations or by the free market, which also results in prices close to cost of supply. The resultant price could be substantially lower than the price of the alternative fuel.
- In countries with shortages of gas, economic analysis should be based on netback values. Also, in a practical sense, gas prices are either set by the government at levels close to the netback value or determined in a free market that pushes price toward netback value.
- In countries that have sufficient gas only for a limited number of years (the "surplus window"), the price of gas should not be as high as the netback value but higher than the cost of production. In such cases, a depletion premium should be calculated and added to the cost of production to arrive at the economic cost of gas supply.

Other Issues

During economic and financial analysis of upstream gas projects a number of other issues may arise:

- The produced gas may be associated with liquids (condensates, LPG). The costs and benefits should be adjusted if the liquids are separated and sold.
- The produced gas may include a significant amount of ethane. The ethane can be separated and sold as feedstock to petrochemical plants.

Table 9.2 Economic and Financial Analysis of Upstream Gas Projects

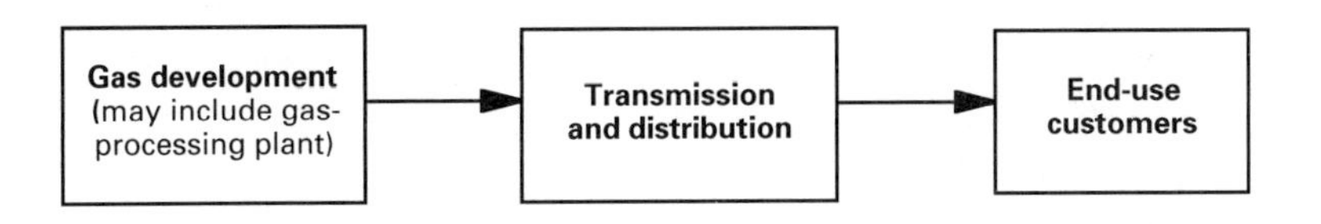

	Financial Costs ($ million)			Financial Benefits ($ million)			
Year	Capital	O&M	Total	Production (BCF)	Price ($/MCF)	Revenues ($ million)	Net Benefits Stream ($ million)
1	15	3	18	0	0	0	–18
2	115	10	125	0	0	0	–125
3	180	30	210	0	0	0	–210
4	120	50	170	0	0	0	–170
5	35	50	85	85	2.1	178.5	93.5
6	60	50	110	90	2.1	189	79
7	48	50	98	110	2.1	231	133
8	14	50	64	105	2.1	220.5	156.5
9	25	52	77	95	2.1	199.5	122.5
10	40	52	92	95	2.1	199.5	107.5
11	32	56	88	95	2.1	199.5	111.5
12	30	52	82	95	2.1	199.5	117.5
13	30	42	72	95	2.1	199.5	127.5
14	0	40	40	95	2.1	199.5	159.5
15	0	40	40	95	2.1	199.5	159.5
16	0	30	30	95	2.1	199.5	169.5
17	0	25	25	95	2.1	199.5	174.5
18	0	20	20	95	2.1	199.5	179.5
19	0	20	20	95	2.1	199.5	179.5
20	0	20	20	95	2.1	199.5	179.5
PV@10%	468.13	307.82	775.95	510.98		1073.06	297.11

Financial IRR = 18% Economic IRR = 37%

Notes: For financial analysis, the capital and O&M costs are based on projected expenses of the company and include taxes and duties. The gas price is the net price to be received by the company.

For economic analysis, capital and O&M costs exclude taxes and duties (estimated at 5% of total cost). The gas price (estimated at $3.6/MCF) is based on the netback value of gas. The choice of gas price for economic analysis would depend on whether the country has a surplus or shortage of gas. This case is adopted from a project in Thailand, where the country has a shortage of gas, and therefore, netback value is an appropriate measure of economic benefit of using gas.

PV = Present Value
BCF = billion cubic feet
MCF = thousand cubic feet

In such a situation, costs and benefits should be adjusted accordingly.

- In calculations of netback values, components of the price of the alternative fuel should be examined carefully. In most developing countries, fuel prices include a variety of taxes. The financial analysis would be based on out-of-pocket cost to the fuel consumer. The economic analysis is based on the cost to the economy and therefore should take prices that exclude taxes and duties.
- In situations where gas is used in several sectors (for example, power, industrial, and residential), separate netback values have to be calculated for each type of use. Economic analysis should weigh each netback value by the amount of gas consumption in the corresponding use.

GAS TRANSMISSION AND DISTRIBUTION

Economic and financial analyses of gas transmission and distribution (T&D) projects must answer three important questions. First, can the new investment be considered as a separate facility, or would it be intertwined with the existing system? Second, can the T&D company own (buy and sell) gas, or would it function as only a carrier of gas? Third, can the economic rate of return be calculated satisfactorily?

Separate versus Intertwined Facilities

Clearly, economic and financial analysis of a T&D project is more straightforward when the proposed investment deals with independent facilities. This would be the case for projects that aim at building new supply systems. For such projects, cost estimates are based on international norms for capital and O&M expenditures. Project benefits are assessed based on the price margin or the throughput tariff of the proposed system.

A new investment intertwined with existing facilities requires an incremental analysis. This involves examining all of the facilities and comparing the gas delivery system with and without the proposed investment. The difference could be only in the volume of the delivered gas or in the reliability of the system, maintenance costs, and so on. All the differences have to be identified, estimated, and incorporated in costs and benefits of the new investment.

Ownership versus Carrier Function

In most developing countries, T&D companies are permitted to own the gas they carry. They buy the gas from producers, deliver and sell it to consumers. Thus, in analyzing project costs and benefits, the cost of purchasing the gas is included in the operations costs. The revenue from selling the gas constitutes the project benefits. If the company is selling to various customer groups, then the average tariff for each group and the relative volumes of gas supply to each group are used in assessing the project benefits.

In countries with significant private sector participation in the gas industry, transmission companies may be barred from owning the gas they carry. Instead, they are required as common carriers to carry the gas for others at a regulated tariff. This arrangement is meant to prevent the transmission company from exercising monopoly power in the gas market. The throughput tariff is supposed to cover capital and operating costs of the pipeline company. The cost and benefit analysis is rather straightforward because project revenues can be simply calculated based on throughput volumes and tariffs.

Calculation of Economic Rate of Return

For most T&D projects, estimates of a financial rate of return can be based on T&D costs and revenues. However, calculation of an economic rate of return is not feasible unless the entire gas chain is considered. Thus, economic analyses often view investments in gas production and T&D as an integrated undertaking. Separate estimation of the economic return for the T&D project is not possible because the benefit due to T&D cannot be isolated from the total benefit. For example, if the cost of gas production is $1.00/MMBTU and cost of T&D is $0.80/MMBTU, the total cost is $1.80/MMBTU. If the consumer receives a benefit of $4.00/MMBTU from using this gas, the net economic benefit is $2.20/MMBTU. However, there is no way to say how much of this benefit is related to T&D and how much to production and development. In effect, the benefit is brought about by the whole chain.

An obvious exception is a country that imports natural gas. In such a case, the price of imported gas is a clear indication of the cost of gas to the economy. The difference between this cost and the netback value represents the benefit that the T&D project would provide by carrying the gas to consumers.

PETROLEUM REFINING

Economic and financial analyses of petroleum refining should consider that costs include investment and O&M costs as well as the cost of purchasing crude oil and that revenues are based on the wholesale prices of petroleum products. The most important issue in the above analyses is the forecast of relevant crude and product prices.

Table 9.3
Financial Analysis of a Gas Transmission Project

Year	Capital cost ($ million)	O&M cost ($ million)	Volume of gas (BCF*)	Purchase price of gas ($/MCF)	Cost of gas purchase ($ million)	Sale price of gas ($/MCF)	Sales revenue ($ million)	Net benefit ($ million)
1	70	0	0	0	0	0	0	–70
2	160	0	0	0	0	0	0	–160
3	60	10	60	2.00	120	2.50	150.00	–40
4	0	10	80	2.06	165	2.58	206.00	31
5	0	10	95	2.12	202	2.65	251.96	40
6	0	10	110	2.19	240	2.73	300.50	50
7	0	10	95	2.25	214	2.81	267.31	43
8	0	10	95	2.32	220	2.90	275.33	45
9	0	10	95	2.39	227	2.99	283.59	47
10	0	10	95	2.46	234	3.07	292.10	48
11	0	10	95	2.53	241	3.17	300.86	50
12	0	10	95	2.61	248	3.26	309.88	52
13	0	10	95	2.69	255	3.36	319.18	54
14	0	10	95	2.77	263	3.46	328.76	56
15	0	10	95	2.85	271	3.56	338.62	58
16	0	10	95	2.94	279	3.67	348.78	60
17	0	10	95	3.03	287	3.78	359.24	62
18	0	10	95	3.12	296	3.89	370.02	64
19	0	10	95	3.21	305	4.01	381.12	66
20	0	10	95	3.31	314	4.13	392.55	69
PV@10%	240.95	67.78	615.84		1501.15		1876.44	67

IRR =13%

Note: The above analysis is based on a case involving construction of 175 km of 32-inch and 160 km of 24-inch pipelines to transport gas from upstream to the point of distribution (city gate).

BCF = billion cubic feet.

The Relevant Crude Oil Price

The problem of forecasting the crude oil price relates to controversies about international price forecasts and possible distortions between domestic and international prices. Where crude oil is imported, financial and, particularly, economic analyses are simplified. The economic price would be based on import parity, and the financial price would include taxes and duties. However, if crude oil is produced domestically, the economic price assessment is based on export parity, but the financial price would be difficult to forecast because it could vary depending on the market power of the refiner and the upstream producers. For these reasons, most financiers prefer to see that the refiner has a firm contract for crude oil supply if the source is domestic.

The Relevant Product Prices

Forecasting product prices is even more complicated than forecasting the crude oil price. First, there are no systematic forecasts for international product prices. The relationship between the crude oil price and product prices has been unstable over time and across countries and markets. Most attempts to formulate and forecast this relationship have been abandoned for lack of success.

Second, in most developing countries domestic product prices diverge from international spot prices of these products. Thus, forecasting product prices in a developing country requires an intimate understanding of the relevant laws, regulations, and government policy, as well as the demand-supply patterns of the product market. Except for "telling" refineries, which operate on a cost-of-service basis, project operating margins will be highly cyclical. Cash shortfalls may affect debt service in cycle troughs, and therefore various mitigation measures may be necessary, such as high initial equity, shareholders' support, netback pricing of captive feedstocks, special reserves, and dividend clawbacks.

Other Issues

Economic and financial analyses of refining projects should further consider the following:

- Forecast of the products mix is an important determinant of project benefits. However, the products mix can vary depending on the configuration of the refinery and market demand. Indeed, the products mix from a refinery will be different in various seasons because of changing demand. These variations must be assessed on the basis of

Figure 9.6
Financial Analysis of Petroleum Refining

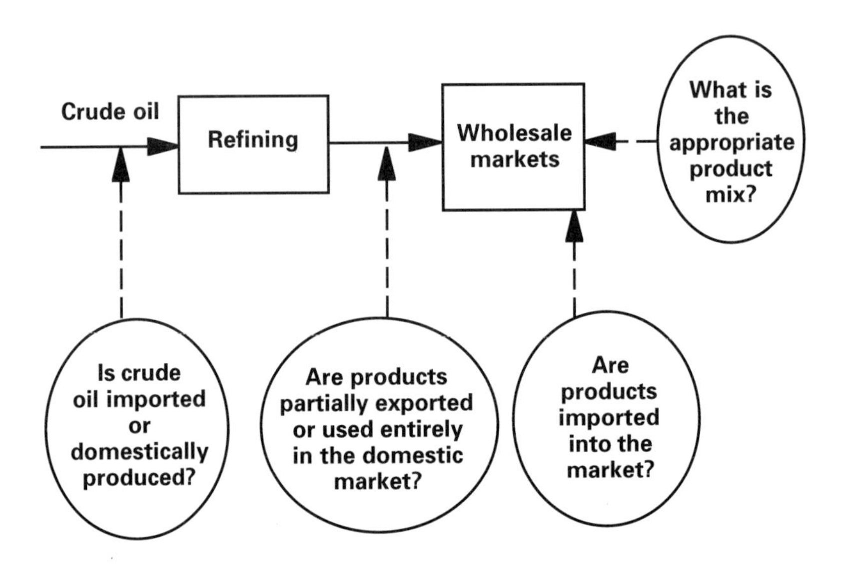

historical data and cross-country observations, and the forecast must take into account foreseeable changes and process obsolescence.

- Many proposed refinery investments are aimed at rehabilitating and upgrading existing refineries rather than building new facilities. The economic and financial analyses should be based on a comparison of costs and outputs with and without the proposed investment.

POWER GENERATION

Economic and financial analysis of the power sector usually includes a least-cost analysis of power generating capacity as well as a calculation of economic and financial ratios.

Table 9.4
Financial Analysis of a Petroleum Refining Project

Year	Capital costs ($ million)	O&M costs[a] ($ million)	Volume of crude oil (b/d)	Price of crude oil ($/bl)	Cost of crude oil[b] ($ million)	Product sales[c] ($ million)	Net Revenue ($ million)
1	2	0	0	0	0	0	–2
2	35	0	0	0	0	0	–35
3	45	0	0	0	0	0	–45
4	42	6	500	20.00	3	11	–40
5	0	14	2500	20.60	17	42	11
6	0	14	2000	21.22	14	43	15
7	0	14	2500	21.85	18	56	24
8	0	14	2800	22.51	21	64	29
9	0	0	2800	23.19	21	65	44
10	0	0	3000	23.88	24	72	48
11	0	0	3000	24.60	24	74	50
12	0	0	3000	25.34	25	76	51
13	0	0	3000	26.10	26	78	52
14	0	0	3000	26.88	27	81	54
15	0	0	3000	27.68	27	83	56
PV@10%	93.24	34.41	12340.51		95.45	285.73	62.63

IRR =18%

Notes:
[a] O&M costs include the cost of fuel used by the refinery.
[b] 330 days/year.
[c] Based on the yield of various products (gasoline, jet fuel, gasoil, fuel oil, and butane) from the refinery and the market prices of these products.

b/d = barrels per day.
PV = present value.

Figure 9.7 Power Generation: The Framework of Least-Cost Analysis

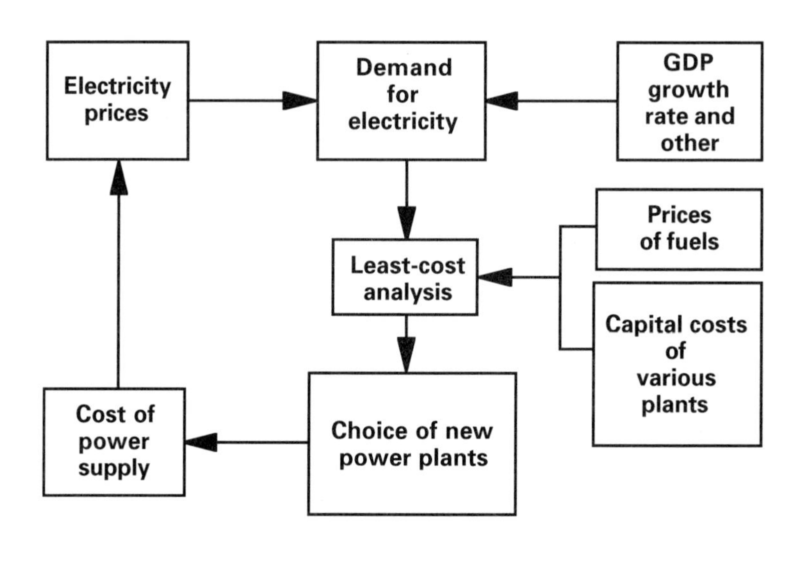

The objective of least-cost analysis is to arrive at a composition of generating capacity that would minimize the overall cost of power supply while satisfying demand within local technical and environmental constraints. The core computations underlying least-cost analysis use computer packages such as the WIEN Automatic System Planning Package (WASP). In these computations, the algorithm takes account of growth in demand; shape of the load curve (variations of demand over 24 hours, days of the week, seasons, and so on); costs of various fuels; technical constraints in loading various plants; and so on. The model then calculates the amount and type of generating capacity that should be added to the system over the planning period.

Although least-cost computational packages facilitate the analysis substantially, it is important to ensure a broad perspective by considering significant links and by using strategic judgment about the results.

Table 9.5 Calculation of Average Levelized Cost

Type of plant = coal-based steam plant
Capacity = 300 MW
Cost = $1,100/kW
Plant cost= $330 million
Plant life = 30 years
Plant load factor = 75%

Efficiency = 34%
Heatrate = 9,914 BTU/kWh
Coal Price = $52/ton
Coal heat content = 26 MMBTU/ton
O&M cost = $16/kW per year + 0.3¢/kWh

Year	Capital Cost ($ million)	Fuel Cost ($ million)	Other O&M Costs ($ million)	Total Cost ($ million)	Power Generation (GWh)
1	33	0	0	33	0
2	66	0	0	66	0
3	99	0	0	99	0
4	99	0	0	99	0
5	33	39.08	10.71	82.79	1971
6	0	39.08	10.71	49.79	1971
7	0	39.08	10.71	49.79	1971
8	0	39.08	10.71	49.79	1971
9	0	39.08	10.71	49.79	1971
10	0	39.08	10.71	49.79	1971
11	0	39.08	10.71	49.79	1971
12	0	39.08	10.71	49.79	1971
13	0	39.08	10.71	49.79	1971
14	0	39.08	10.71	49.79	1971
15	0	39.08	10.71	49.79	1971
16	0	39.08	10.71	49.79	1971
17	0	39.08	10.71	49.79	1971
18	0	39.08	10.71	49.79	1971
19	0	39.08	10.71	49.79	1971
20	0	39.08	10.71	49.79	1971
21	0	39.08	10.71	49.79	1971
22	0	39.08	10.71	49.79	1971
23	0	39.08	10.71	49.79	1971
24	0	39.08	10.71	49.79	1971
25	0	39.08	10.71	49.79	1971
26	0	39.08	10.71	49.79	1971
27	0	39.08	10.71	49.79	1971
28	0	39.08	10.71	49.79	1971
29	0	39.08	10.71	49.79	1971
30	0	39.08	10.71	49.79	1971
31	0	39.08	10.71	49.79	1971
32	0	39.08	10.71	49.79	1971
33	0	39.08	10.71	49.79	1971
PV@10%	247.03	250.10	68.54	565.67	12,613.55

Average capital cost = 247.03/12613.55 = $0.02/kWh = 2.0¢/kWh.
Average fuel cost = 250.10/12613.55 = $0.02/kWh = 2.0¢/kWh.
Average O&M cost = 68.54/12613.55 = $0.005/kWh = 0.5¢/kWh.
Average cost of power generation = 565.67/12613.55 = $0.045/kWh = 4.5¢/kWh.

Table 9.6 Comparison of Average Levelized Costs for Coal, Oil and Gas Power Plants

	Coal-based steam plant	Oil-based steam plant	Gas-based combined-cycle
Capital cost	2.0¢/kWh	1.6¢/kWh	1.1¢/kWh
Fuel cost	2.0¢/kWh	2.9¢/kWh	2.1¢/kWh
O&M cost	0.5¢/kWh	0.3¢/kWh	0.3¢/kWh
Average cost of power generation	4.5¢/kWh	4.8¢/kWh	3.5¢/kWh
Assumptions			
Capacity (MW)	300	300	300
Cost ($/kW)	1,100	900	650
Plant cost ($ million)	330	195	270
Plant life (years)	30	2	25
Efficiency (%)	34	36	48
Heat rate (BTU/kWh)	9,914	9,630	7,145
Fuel price	$52/ton	$18/bl	$3/MCF
Heat Content	26 MMBTU/ton	6 MMBTU/bl	1 MMBTU/MCF
O&M cost	$16/kW per year +0.3¢/kWh	$11.50/kW per year +0.175¢/kWh	$8/kW per year +0.2¢/kWh

To take account of significant links, it is necessary to go through several iterations to (1) assess the cost of power supply, (2) estimate the future electricity price, (3) examine the impact of the price on electricity demand, and (4) feed back the revised demand forecast into the least-cost analysis.

The Relevance of Average Levelized Cost

In the economic and financial analysis of power generation, it is helpful to calculate the average levelized cost of various plant options. This measure is also referred to as the average discounted cost or, under certain conditions, average incremental cost. It provides a simple measure of the average capital and fuel costs for each type of generating plant. The calculation is as follows:

Average levelized cost = (PV of cost stream)/(PV of output stream),
where PV is present value.

Table 9.7 Power Generation: Calculation of Economic Rate of Return

	Investment ($ m)				Operation ($ m)					Total Cost ($ m)	Incremtl. sales (GWh)	Rev. ($ m)	Net Benefit ($ m)
Year	Gen.	Trans.	Dist.	Total	Gen.	Trans.	Dist.	Fuel	Total				
1	15.9	17.5	20	53.4	0	0	0.2	0	0.2	53.6	0	0	–53.6
2	47.6	52.6	20	120.20	0	0	0.5	0	0.5	120.7	0	0	–120.7
3	15.9	35.1	25	76	0.1	0	0.8	1.1	2	78	22.7	6.4	–71.6
4	20.9	17	26	63.9	1.4	1.6	1.1	13.4	17.5	81.4	273	76.4	–5
5	62.8	0	36	98.8	1.5	1.6	1.5	15.7	20.3	119.1	357.8	100.2	–18.9
6	20.9	0	38	58.9	1.7	1.6	2	19.7	25	83.9	357.8	100.2	16.3
7	0	0	35	35	4.5	1.6	2.4	20.3	28.8	63.8	535.9	150.1	86.3
8	0	0	0	0	4.5	1.6	2.4	21.5	30	30	535.9	150.1	120.1
9	0	0	0	0	4.5	1.6	2.4	22.6	31.1	31.1	535.9	150.1	119
10	0	0	0	0	4.5	1.6	2.4	23.7	32.2	32.2	535.9	150.1	117.9
11	0	0	0	0	4.5	1.6	2.4	24.8	33.3	33.3	535.9	150.1	116.8
12	0	0	0	0	4.5	1.6	2.4	26	34.5	34.5	535.9	150.1	115.6
13	0	0	0	0	4.5	1.6	2.4	27.7	36.2	36.2	535.9	150.1	113.9
14	0	0	0	0	4.5	1.6	2.4	28.8	37.3	37.3	535.9	150.1	112.8
15	0	0	0	0	4.5	1.6	2.4	30.5	39	39	535.9	150.1	111.1
16	0	0	0	0	4.5	1.6	2.4	31.6	40.1	40.1	535.9	150.1	110
17	0	0	0	0	4.5	1.6	2.4	33.3	41.8	41.8	535.9	150.1	108.3
18	0	0	0	0	4.5	1.6	2.4	35	43.5	43.5	535.9	150.1	106.6
19	0	0	0	0	4.5	1.6	2.4	36.7	45.2	45.2	535.9	150.1	104.9
20	0	0	0	0	4.5	1.6	2.4	38.4	46.9	46.9	535.9	150.1	103.2
21	0	0	0	0	4.5	1.6	2.4	40.6	49.1	49.1	535.9	150.1	101
22	0	0	0	0	4.5	1.6	2.4	40.9	49.4	49.4	535.9	150.1	100.7
23	0	0	0	0	4.5	1.6	2.4	44.6	53.1	53.1	535.9	150.1	97
24	0	0	0	0	4.5	1.6	2.4	47	55.5	55.5	535.9	150.1	94.6
25	0	0	0	0	4.5	1.6	2.4	47	55.5	55.5	535.9	150.1	94.6
26	0	0	0	0	4.5	1.6	2.4	47	55.5	55.5	535.9	150.1	94.6
27	0	0	0	0	4.5	1.6	2.4	47	55.5	55.5	535.9	150.1	94.6
28	0	0	0	0	4.5	1.6	2.4	47	55.5	55.5	535.9	150.1	94.6
PV @10%	130.81	97.36	133.02	361.18	25.20	10.91	15.89	175.76	228.77	589.95	3281.06	918.96	329.01

IRR = 20%

Notes: Based on a project in Malaysia. The power plant is a 90-MW gas-based combined-cycle plant. The generated power is transmitted through a 185-km, 275 kilovolt (kv) transmission line to the corresponding consumption center. Additional investments in transmission and distribution are included. Fuel price is based on the economic cost of natural gas. Price of electricity is the weighted average of tariffs and is assumed to stay at the same level in real terms.
Gen. = generation; Trans. = transmission; Dist. = distibution; Incremtl. = Incremental; Rev. = Revenue.

Calculation of IRR

Calculation of *financial* IRR is based on costs and revenues the project actually will face. Capital and O&M cost estimates are based on international norms and local conditions. The fuel cost is more difficult to forecast because of the lack of generally accepted forecasts of international prices and because of possible distortions in fuel prices in local markets. The risk associated with fuel costs will be minimized if project sponsors can secure a fuel supply contract with a clear price formula. In some build-operate-transfer (BOT) projects, sponsors have managed to shift the risk of fuel cost to the national power company through an arrangement in which the power company provides fuel to the BOT sponsors and receives back generated power. Under these arrangements, BOT functions as a processing unit—receiving fuel and giving back the generated electricity.

The second difficulty in calculating IRR is in forecasting the electricity price. If the project is being built by the national power company, the electricity price estimate should be based on power tariffs and sales to various customer groups. In the case of BOT suppliers, the price risk is minimized through a take-or-pay contract in which project sponsors receive assurances from the power grid system (normally a public utility) that the generated power will be bought based on a clear price formula.

Calculation of *economic* IRR is based on costs and benefits of the project to the economy rather than the company. The cost figures should be adjusted to exclude taxes and duties and to compensate for other distortions—in exchange rates, fuel prices, and wages. The benefit stream should be based on the "true" economic benefit of electricity to consumers. Two difficulties arise in assessing the true benefit. First, power tariffs normally underestimate these benefits. An attempt is sometimes made to assess the true benefit through surveys of various customer groups. These surveys examine the behavior of the customers and estimate the cost imposed on each customer group by interruption or shortage of power. Second, power tariffs are economically meaningful only at the point of consumption. Thus, economic IRR is normally calculated for the entire incremental investment—that is, investment in power generation plus the associated investments required to expand the T&D system to carry the additional electricity to the consumer.

POWER TRANSMISSION AND DISTRIBUTION

Economic and financial analysis of T&D projects is normally carried out on the "with and without" basis. In the event that the transmission line is specifically used to carry power from one to another point, the "without" case would involve alternative forms of energy transport. For example, in a country where gas reserves are far from the main centers of power consumption, one option is to build a power plant near the gas field and a transmission line to move the generated power to the consumption center. The alternative would be to construct the power plant near the consumption center and a pipeline to move the gas from the field to the power plant. Thus, the alternative to the power transmission line is a gas pipeline. Economic evaluation of the transmission project is therefore carried out as follows:

- The capital and operating costs of the transmission line are estimated on the bases of international norms and local conditions.
- The benefit stream would be based on the cost of the alternative—the capital and O&M costs of the pipeline—as well as other differences; for example, the power plant may cost more to build if it is near a city. It may also include the value of serving new customers along the transmission line (electric versus gas).

More often, T&D investments are aimed at expanding or reinforcing existing networks. These investments result in reduction of technical losses, improved system reliability, and increased capacity. The comparison of "with and without" cases would require quantification of these benefits. Technical losses for the grid (with and without proposed investments) can be assessed with network simulation packages. The reduction in losses is then valued based on the average power tariff. The value of the improved system reliability is difficult to assess. It is normally estimated on the basis of the change in the probability of power outage and the cost of outage to consumers. The economic value of increased capacity is assessed in relation to the power tariffs (capacity and energy charge) minus the cost of power generation.

The economic and financial analyses of power T&D projects may raise the following additional issues:

- As a special case, the transmission line may be entirely associated with a power plant; that is, it is built specifically to transmit power from a plant to the grid. The economic and financial analysis is then performed for the entire generation and transmission scheme.

- Rural electrification is a special T&D project that aims at taking the power from the grid to a specific consumption center. Such a project can be evaluated as an independent investment. The project benefits will be assessed on the basis of the cost of energy to the rural community in the absence of access to electricity.

10

STRUCTURING THE FINANCING PACKAGE

Although state-owned and privately owned energy companies may be distinguished sharply in theoretical terms, they converge surprisingly in the way they seek to finance energy projects in developing countries. State-owned oil, gas, and power companies, which traditionally financed their projects through government budgets or official government-sponsored borrowing, have now turned to commercial sources of finance such as commercial banks, private bond placements, and even stock markets. On the other hand, private investors that are supposed to bring in commercial funds often seek ways to use official sources of finance with some type of government sponsorship. This occurs most readily where private sponsors form joint ventures with state entities.

In a similar way, the distinction between corporate (balance sheet) financing and project financing is becoming less pronounced in developing countries. On the one hand, pure corporate financing is becoming rare. Most corporations do not want to keep project debts on their balance sheets for the entire periods of repayment; they try to move the debt off their balance sheets as soon as possible, particularly after the project construction phases. On the other hand, pure project-based financing has turned out to be very difficult for large energy projects in developing countries. Sponsors can mobilize equity and debt financing more easily when they draw on the creditworthiness of parent companies. Therefore, corporate financing and project-based financing are converging toward an arrangement of limited corporate responsibilities in the form of limited-recourse or other types of sponsorship.

As a result, project preparation and the associated analysis are becoming similar whether the project comes under state or private ownership, or whether the project is supposed to be financed on a corporate or project basis. However, the extent of the required risk mitigation measures varies with the creditworthiness of involved parties and the role of the state in ownership, construction, and operation of the plant.

Figure 10.1 Ownership and Financing Structures

Determining the ownership structure is extensively intertwined with the design of the financing structure and security package. These items will be continuously modified through an iterative process right until the end of the preparation process and sometimes through the implementation phase.

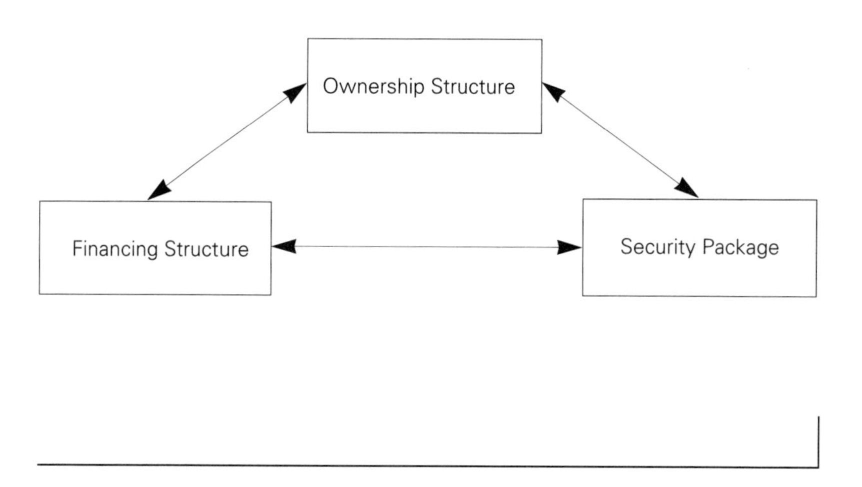

DETERMINING THE OWNERSHIP STRUCTURE

The ownership structure is determined in two steps: conceptual and technical. The conceptual stage is the time for decisions about the role of various participants, particularly state versus private and foreign versus local relationships. At the technical stage the legal structure is devised based on a combination of financial, accounting, and tax considerations.

Ownership structures fall into three broad categories:
Complete state ownership, which should be the most convenient structure from the point of view of financing; a project undertaken by a state entity

can be funded from both public and private sources of finance. However, complete state ownership has certain drawbacks. First, concerns may arise about the efficiency of most state-owned energy entities of developing countries. Such concerns are translated into perceptions of risk regarding construction costs and schedule and regarding plant operation and maintenance. Second, most multilateral and bilateral financiers discourage full state ownership of energy projects and emphasize private sector participation. As a result, many financiers have some hesitation about funding projects fully owned by the state. The hesitation varies with the host-country circumstances and the type of energy project. For most developing countries, the state is not expected to have full ownership of upstream oil and gas projects or of downstream activities such as refining and product distribution. On the other hand, state ownership is justified in the case of gas infrastructure and power transmission and distribution (T&D). In regard to power generation, private power producers are accepted worldwide as an alternative to electricity generation by state utilities. Nevertheless, there is still substantial room for generation by state utilities.

Complete private ownership by a special-purpose company (project company) or credible corporations. The international community has demonstrated support for this type of ownership structure in most segments of the energy sector, particularly upstream and downstream petroleum investments. However, several drawbacks are apparent. For gas infrastructure and power transmission and distribution, regulations must deal with the natural monopoly characteristics of these facilities. Even in the case of power generation, where private ownership is common and accepted, numerous regulatory matters must be resolved. The challenges of undertaking a wholly private project are more manageable when project sponsors are reputable companies willing to accept formal obligations for efficient construction and operation of project facilities. In addition, direct guarantees by such companies facilitate debt financing. Overall success also depends on the composition of partnership. A composition likely to be successful would involve powerful local partners and international firms with complementary lines of business—for example, an engineering firm, an equipment supplier, and a construction company.

Joint ventures between state and private parties. This type of ownership structure is becoming increasingly popular because it provides a very flexible framework for sharing project risks and a wide range of options for equity and debt financing. The most advantageous venture would be between a credible large corporation and an efficient state entity.

It is neither feasible nor desirable to fix the ownership structure at the early stages of project preparation. The ownership structure is determined through an iterative process in which the initial sponsors assess the costs and benefits of involving various sources of equity, finance, and risk management. The optimal structure would differ depending on the type of project, the country, and the sponsors' credibility and objectives. However, common objectives normally pursued in devising ownership and financing structures are (1) to utilize state direct and indirect support as much as possible while minimizing state intervention in the managerial aspects of project construction and operation and (2) generally to allocate risks, contributions, and benefits fairly among the interested parties, with due consideration for their respective abilities.

Overview 10.1 Determining the Ownership Structure

The ownership structure of a project is determined in two steps: conceptual and technical. Important conceptual stage considerations are

- Complete state ownership formally provides access to a variety of sources of funds, but in practice most official sources of funds hesitate to support full state ownership, particularly for upstream oil and gas and downstream oil projects.
- Complete private ownership is supported by almost all players in the energy sector but cannot be easily formulated in the case of gas infrastructure and power transmission and distribution (T&D).
- Joint ventures between public and private sector entities are becoming very popular because of the flexibility they provide for sharing risks and mobilizing debt and equity finance.

At the technical stage, the ownership structure (for example, partnership, joint venture, or single stock company) is determined on the basis of the sponsors' tax, accounting, and risk-sharing objectives.Technical design of ownership structure, like conceptual design, is arranged in conjunction with the financing and security structures and deals with the type of company (joint venture, partnership, or single-purpose stock company); place of incorporation or establishment; and so on. These arrangements should be devised to

- Optimize the tax position of the sponsors, taking into account both host- and home-country tax rules.
- Achieve the accounting treatment and risk-sharing objectives of the sponsors.
- Achieve the desired risk-sharing objectives of the sponsors.
- Maximize the level of nonrecourse financing at lowest cost.
- Provide protection for individual sponsors against default by other sponsors.

The ownership structure should also include a choice between whether each partner is to be individually responsible for financing, as is typical in oil projects, or whether finance is to be raised for the venture as a whole ("common financing") backed by several undertakings of the sponsors and appropriate interpartner default clauses. Common financing is normally used in the power sector.

IDENTIFYING THE SOURCES OF FINANCE

Part II of this book describes the background, methods, and conditions of accessing equity and debt finance from multilateral, bilateral, and commercial sources. For most energy projects, sponsors start their investigation of equity and debt financing by considering the entire menu of options. Nevertheless, identifying sources appropriate to the project requires a systematic procedure, as explained below.

Assume a Flexible Ownership Structure

This is normally done by envisaging participation by foreign as well as local private and state partners.

Examine the Government Interest

The government interest in the project should be examined at several junctures with two distinct objectives. First, would government provide or sponsor equity or debt financing under any circumstance? Second, would the government provide administrative support and facility in case it does not wish to be involved in financing? Even a simple declaration of "national interest" from a "responsible" government may help, especially in bilateral negotiations.

Assess the Receptivity of the Multilaterals

These agencies may not be interested in certain projects or countries regardless of the sponsorship for the project. If the multilaterals do indicate interest, project sponsors should consider an ownership structure that accommodates their requirements.

Approach the Bilaterals

Sponsors should approach bilateral aid agencies of their own countries with the objectives of getting support for project preparation and securing finance in the form of loans or loan guarantees. In addition, they should approach bilateral agencies of other countries with well-known suppliers of goods and services needed for the project. Some bilaterals provide investment financing in return for a commitment that the output of certain projects would be imported into their countries. For example, if the output of an oil or gas project can be allocated for export to Japan, project sponsors may be able to obtain loan or loan guarantees from the Export-Import Bank of Japan (JExim).

Ask Equipment Suppliers to Mobilize Finance

As soon as the overall scope and design of the project are known, project sponsors can identify the most important sources of equipment supply and construction and engineering services. Quite often some of these businesses are reserved for project sponsors or their affiliated companies. Still, significant components of the project may be acquired from international manufacturers and contractors. Large equipment-supply companies, including trading houses, may contribute funds from their own resources. More importantly, they have substantial influence with export credit agencies of their own countries and can mobilize loans or guarantees from them.

Use the IFC or Regional Development Banks to Mobilize Commercial Funds

Large corporations, such as major oil companies, are normally able to access commercial sources of equity and debt finance directly, but they may not wish to limit the use of their balance sheet for a particular project. Other, "weaker" project sponsors have difficulty doing so on their own.

Overview 10.2 Identifying Sources of Finance

In identifying the sources of finance, sponsors should

- Assume a flexible ownership structure.
- Examine the interest of the host government in the project.
- Assess the receptivity of multilateral agencies.
- Approach bilaterals of their own countries as well as countries that can supply equipment and services.
- Ask equipment suppliers to mobilize finance.
- Use International Finance Corporation (IFC) or regional development banks to mobilize domestic and international commercial funds.
- Review the availability of specialized energy funds.

Institutions such as the International Finance Corporation (IFC) and the regional multilaterals have experience in quickly assessing the applicability and viability of various arrangements. They can assist by providing and mobilizing equity from institutional investors. They can also structure syndicated loans and mobilize borrowing on bond markets. However, sponsors should be aware that the IFC, like other financial advisors, charges for the cost of its advisory services.

Review the Availability of Specialized Energy Funds

These funds, either government-sponsored or private, may or may not be available for the project and the country of concern. Nevertheless, if available, these funds can facilitate access to both equity and debt financing.

Survey Domestic Investors and Lenders

Domestic investment and lending are becoming important sources of funding for energy projects in developing countries that have advanced capital markets. In particular, two categories of financiers should be approached. First, rich individuals willing to take calculated risks by participating in well-designed ventures can often facilitate preparatory work of the project because of their familiarity and influence in the bureaucracy and other local affairs. Second, institutional investors and, occasionally, local banks can be persuaded to participate when credible international companies and institutions are involved. Demonstration of local and international support may allow international capital markets to be tapped, especially after project completion (see chapter 7).

PREPARING THE SECURITY PACKAGE

The security package comprises key agreements, contracts, and government undertakings aimed at reducing lenders' and investors' risks by establishing legally binding obligations and procedures. Some documents included in the package represent standard paperwork prepared for a project regardless of who the lenders and investors are. Examples are land use and operating license, construction permit, import license, corporate documents, trust agreements, concession agreement, production-sharing agreement, and offtake agreement.

The security package also includes various guarantees that need to be designed to mitigate project risks. These vary depending on whether the major portion of debt is financed by commercial or noncommercial lenders.

From a lender's point of view, three questions need fully satisfactory answers. First, can the project be constructed and commissioned within the planned schedule and budget? Second, can the project generate the projected net revenue? Third, can the net revenue be allocated and paid back to the lenders and investors according to the project agreement? Associated with these questions, lenders want to know who would be responsible for damages in the event the project fails in any of these areas. The security package should provide the lenders with answers to these questions as well as the means to perfect and enforce their security interest.

Preparation of the security package significantly affects and is affected by the ownership and financing structures. In particular, participation by the government in ownership or participation by multilaterals in financing would reduce concerns about political risk and sometimes increase concerns about commercial risks. The point is that the security package will continue to be modified until the ownership and financing structures are finalized. Preparing an appropriate security package requires a systematic procedure, as explained below.

Identify Sources of Applicable Guarantees

It is not feasible or advisable to ask for commitment from any commercial or official sources of guarantee before deciding about partners and financiers. A guarantor would need to know other important players. In addition, the structure of financing affects the options for procurement of goods and services, which in turn affects the choice of guarantor. Nevertheless, it is possible to compile a list of potential guarantors and inquire about their interest in participating in the proposed project. This information is needed before approaching financiers.

Evaluate Options for Mitigating Political Risks

Although mitigating commercial risks represents the most time-consuming aspect of preparing the security package, the issue of political risk should be addressed at the outset. Most investors and financiers are convinced that commercial risks can be effectively addressed when the time comes, but they also feel that political risks are out of everybody's control. Thus, they do not take a proposal seriously until they receive some comfort that political risks are manageable. Options for mitigating political risks include several types of guarantees as well as the involvement of certain players—for example, partnership with a key state entity or with powerful local individuals and companies.

Formal guarantees against political risk are provided by the host government and by multilateral and bilateral agencies. Often, the question is not about choosing one versus another but about combining them to provide the most comprehensive coverage at the lowest possible cost. Although it is a "partial guarantee," the World Bank's program is a relatively broad guarantee instrument that should be viewed as a good starting point in sponsors' investigations for political risk guarantee. The World Bank also provides leads to other applicable sources of guarantee. Other multilateral institutions—for example, the regional banks—also provide guarantees and assistance in identifying sources of guarantee. Most multilaterals have regular working relationships with the bilateral agencies, know of their country and sectoral preferences, and cooperate with them in processing requests for loans and guarantees.

Investigate Options for Mitigating Commercial Risks

This is certainly the most time-consuming aspect of preparation of the security package. The task has two distinct dimensions. First, sponsors need to reach agreement with the government or certain government entities about some aspects of marketing the project's output. The government's role varies depending on the country and the type of project. However, for gas and power projects the government's role is substantial because the output is primarily bought by a state entity or because prices are regulated by the state. Therefore, project sponsors need to secure take-or-pay or throughput agreements with the state entities. The government would need to guarantee credibility of the state entities or that it will permit a sufficient increase in energy prices. Securing government guarantees and agreements takes a relatively long time to complete, particularly in countries that lack clear precedents.

The second dimension in mitigating commercial risk involves negotiating with contractors, equipment suppliers, fuel suppliers, operating companies, and so on to determine their willingness to compensate for damages if they fail to fulfill their obligations. This is a technically complex process, but it is normally accomplished efficiently because it is driven by commercial incentives. The prerequisite for completion of these negotiations is a firm plan for procurement of goods and services.

Overview 10.3 Preparing the Security Package

- The security package includes some standard documents (such as permits and licenses) and various guarantees or agreements aimed at (1) managing and mitigating risk, (2) enabling the pledge of sponsors' assets, and (3) perfecting and enforcing the security interest.
- Contents of the security package will be modified continuously through the preparation process because they are affected by the ownership and financing structures. In particular, an increased government role in the project may reduce the need for arranging insurance and guarantees against political risk but may also increase the need for guarantees against commercial risks.
- Although mitigation of all risks will have to be considered, the issue of political risk should be addressed at the outset. Most investors and financiers do not take a proposal seriously until they receive some comfort that political risks are manageable. Among various commercial risks, handling the market risk, and the corresponding guarantee instrument such as a take-or-pay contract, are the most challenging. Somewhere between political and commercial risks is the currency exchange risk.

EXAMPLES OF FINANCING STRUCTURES

Although it is not feasible to present standard models of financing structure for energy projects, experience in this area is instructive. Examples discussed in this section are chosen with the objective of presenting various types of arrangements. Together, these examples represent the menu of instruments and the art of selecting the right combination for the prevailing circumstances.

Financing Upstream Oil and Gas Projects

Upstream oil and gas projects in developing countries are normally developed either by the country's national oil company (NOC) or by ventures involving international oil companies (IOCs). Financing for these projects is usually based on internal cash generation and corporate borrowing. This is particularly true for projects developed by large IOCs, which utilize a corporate pool of funds mobilized through a wide range of financial instruments and markets. For many smaller IOCs, corporate borrowing is limited by the strength of their balance sheets. They often borrow funds on nonrecourse or limited-recourse bases. The following examples indicate the types of financing structures for state-owned, private, and public-private/joint-venture projects.

The Sichuan Gas Development Project in China. Review of the Sichuan gas project demonstrates how a state-owned upstream project can be financed. It also indicates how various sources of grants can be utilized to finance project preparation and implementation.

Sichuan is the most populous province in China. Responsibility for gas production, distribution, and transmission lies with the Sichuan Petroleum Administration (SPA), a subsidiary of the China National Petroleum Corporation (CNPC).

Gas production in Sichuan peaked in 1978 and declined thereafter because of aging wells. Remedial work in the early 1990s reversed the decline. A major project was formulated in 1994 to prevent future decline. The project consisted of a plan for development of gas reserves, rehabilitation of old gas wells, and associated expansion and rehabilitation of the transmission system. The total project cost was estimated at $945 million.

Figure 10.2 Financing Structure of the Sichuan Gas Development Project

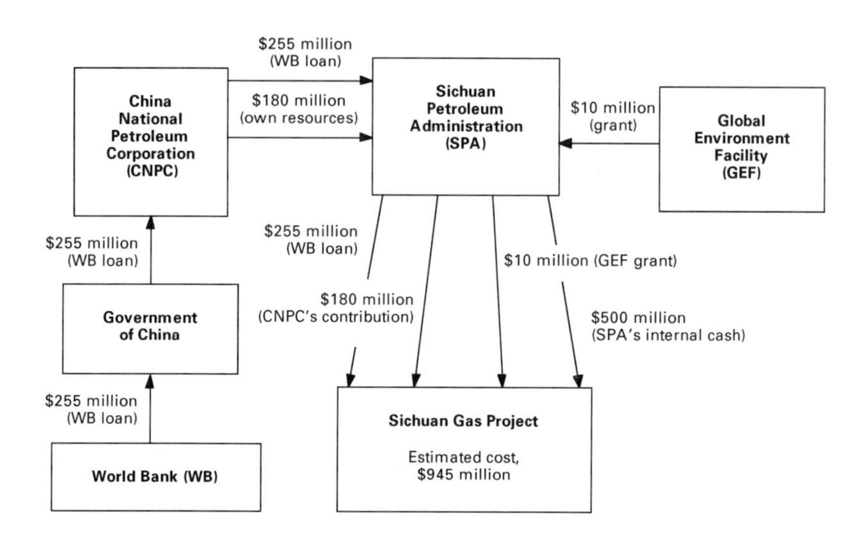

The project financing plan is as follows:

SPA internal cash	$500 million
CNPC internal resources	$180 million
Global Environment Facility (GEF)	$10 million
World Bank	$255 million
Total	$945 million

Although the financing structure is simple, it includes some interesting features. First, preparation of the project was supported by grant funds from Japan, the United Nations Development Programme (UNDP), the GEF, and others. Because of the environmental advantage of natural gas over coal, and because Sichuan is a poor province, the international community had substantial sympathy for the project. Second, project implementation was also partly financed by a GEF grant of $10 million. The justification for GEF grant funding was that rehabilitation of SPA's

T&D system is expected to reduce leaks of natural gas, whose main component, methane, is a potent greenhouse gas.

Bridas Investments in Argentina. Bridas Corporation, a privately held company incorporated in the British Virgin Islands, is engaged (through subsidiaries) in three principal areas of operations: (1) exploration, development, and production of oil and natural gas; (2) gathering, treatment, and processing of gas; and (3) transportation and marketing of oil and gas. More than 80 percent of Bridas' business is in Argentina. The company conducts its activities in Argentina through Bridas SAPIC (BS), a subsidiary. BS had a net worth of $787 million in 1994, when its management decided to mobilize finance for 1995-96 upstream investments totaling $201 million. About half of the capital expenditure was aimed at development and production enhancement of three gas-bearing blocks. Most of the remainder of BS's capital investments was spread over nine oil and gas properties. The investment program included expenditures for surface facilities; a branch pipeline; and other typical development activities such as drilling production wells, workovers, and secondary recovery schemes. The program also included a 15 percent component for exploration as part of its normal reserve replacement effort.

Composition of the investment program is as follows:

Investments in the main three gas blocks	$98 million
Other oil and gas investment	$72 million
Exploration	$31 million
Total	$201 million

Financing for the investment program is largely (65 percent) based on internal cash generation of the company. The rest of the financing was arranged by the IFC, either through IFC's own resources (equity and A loan), or through syndication of loans from commercial banks (B loan).

Equity	
BS's internal cash	$131 million
IFC equity participation	$10 million
Debt	
IFC A loan	$20 million
IFC B loan	$40 million
Total	$201 million

Figure 10.3 Financing Structure of Bridas Investments in Argentina

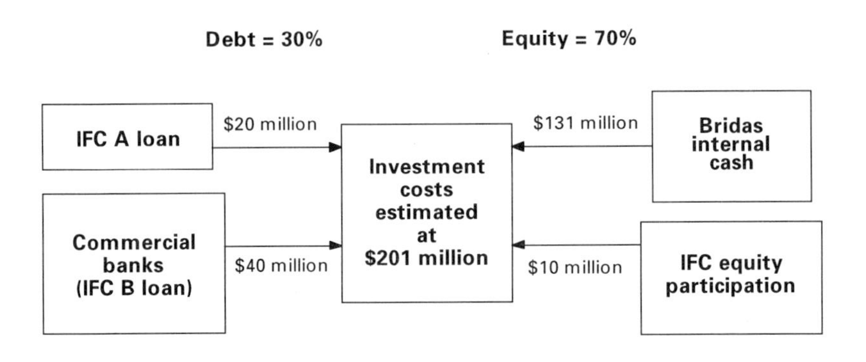

The Cusiana Oil Pipeline in Colombia. The project, commonly referred to as Cusiana, is based on a development plan for two adjacent crude oil reservoirs, the Cusiana and the Cupiagua fields, in Colombia. The exploration licenses for these blocks were initially given to Triton in the early 1980s. British Petroleum (BP) and Total together subsequently farmed in to these blocks, with BP assuming operatorship. Working interests of the three foreign partners during the exploration and appraisal phases were as follows: BP and Total, 38 percent each; Triton, 24 percent. Under a standard provision for contracts in Colombia, Ecopetrol (the Colombian state oil company) exercised its right to take up a 50 percent working interest in the licenses following a formal declaration of commerciality in 1993. The development project was accordingly undertaken by Ecopetrol (50 percent), BP and Total (19 percent each), and Triton (12 percent). Each partner's production rights are subject to an overriding government royalty of 20 percent, reducing the net production shares for BP and Total to 15.2 percent each and for Triton to 9.6 percent. Each foreign partner is subject to the normal corporation tax regime applicable to industrial activity in Colombia.

Two well-known projects are associated with Cusiana. The first, known as Phase 1, is development of Cusiana and Cupiagua fields within the limits of existing pipeline capacity. The second, Phase 2, aims at substantially expanding the pipeline capacity and increasing upstream production capacity to 500 million barrels per day.

The first project, costing about $1 billion, was financed by the partners in proportion to their shares. Each partner took responsibility for mobilizing its own financing. There was no project-based financing except for USExim credits used to fund contributions of two individual partners.

The funding of the pipeline for Phase 2 took a different course and mobilized finance on a project basis. The main reason was a change in government policy in regard to contributing public budget to oil projects. In 1994, the government announced its intention that pipeline infrastructures were to be developed by the private sector rather than under the dominant influence of Ecopetrol. To facilitate this process, the government expressed a preference that the pipeline project be constructed by a joint-stock company in which the foreign partners could, if desired, take shareholdings corresponding to their own transportation requirements. Accordingly, a capital stock company, Ocensa, was created in December 1994 for the purpose of constructing the Oleoducto Central Pipeline. The new pipeline (about 800 km), scheduled for completion in 1997, generally follows the existing Phase 1 pipeline route from the Cusiana and Cupiagua oil fields to the port of Covenas on the Caribbean coast. Crude oil is transported from this port to U.S. Gulf and East Coast ports.

Ocensa's shareholders are Ecopetrol, 25 percent; BP Colombia Pipeline, Ltd. (U.K.), 15.2 percent; Total Pipeline Colombie (France), 15.2 percent; Triton Pipeline Colombia (U.S.), 9.6 percent; Interprovincial Pipelines Enterprises (IPL; Canada), 17.5 percent; and TCPL International (TransCanada), 17.5 percent. BP, Total, and Triton will use their share of the pipeline system to ship their own oil. IPL and TCPL will ship Ecopetrol's oil.

The financing plan structure is 30 percent equity and 70 percent debt financing. The equity portion—$600 million—consists of $500 million paid-in cash by shareholders and $100 million expected in cash from tariffs.

The debt is financed through a variety of methods. The actual borrower is Ocensa, but each oil shipper arranged its own portion of debt, according

Figure 10.4 Financing Structure of the Cusaiana Oil Pipeline in Columbia

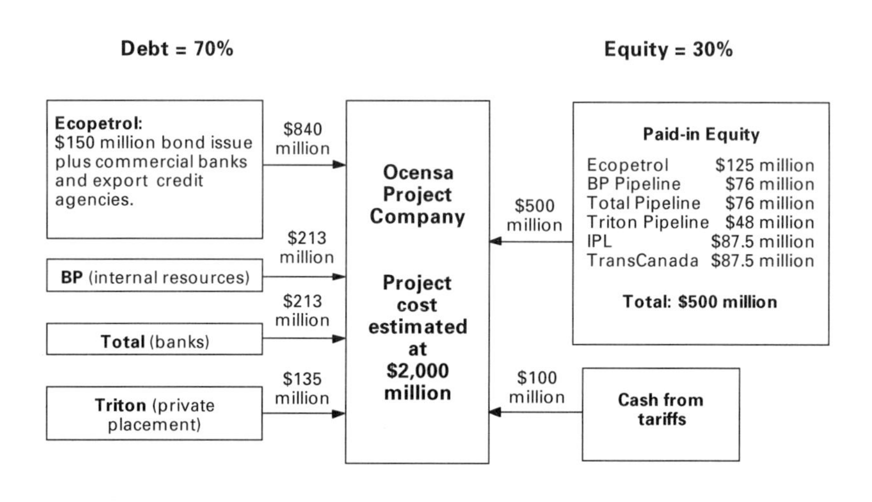

to its interest in the upstream field. Ecopetrol used a combination of bond issues, commercial bank borrowing, and loans from export credit agencies to borrow $840 million. BP continued its practice of funding projects from its corporate resources. Total used a consortium of banks to arrange $210 million in commercial loans. Trinton financed its share of debt through private placement with institutional investors. The borrowing by each company is backed by transportation agreements for shipment of the company's oil through the pipeline system. Ecopetrol's borrowings are backed by a transportation agreement for shipment of Ecopetrol's oil.

Financing Petroleum Refining Projects

Petroleum refineries are built by private companies in most developing countries. Even in countries where a state company owns the refining facilities, new investments are usually funded by commercial sources or a combination of public and commercial sources.

In the case of the Star Refinery project, in Thailand, the government of Thailand invited proposals for construction of additional refineries in 1989, responding to shortages of domestically refined petroleum products. In 1991, Caltex Petroleum Corporation proposed and reached an agreement to construct a refinery in Rayong Province, about 220 km southeast of Bangkok. A project company, Star Petroleum Refining Company (SPRC), was formed in 1992 as a joint venture between CXO Limited and the Petroleum Authority of Thailand (PTT). CXO owns 64 percent and PTT 36 percent. CXO is a joint venture between the Oman Oil Company and the Transport Corporation, a wholly owned subsidiary of Caltex Petroleum Corporation.

The Star refinery will be a 130,000 barrel per day (b/d) deep-conversion refinery designed to maximize production of unleaded gasoline and low-sulfur diesel for Thailand's domestic market.

The project cost is $1.8 billion. The financing plan was based on 30 percent equity and 70 percent debt. The equity portion is provided by project sponsors in accordance with their shares. The debt portion was financed as follows:

IFC A loan	$75 million
IFC B loan	$275 million
JExim supplier's credit	$400 million
Parallel commercial loan	$449 million
Thai revolving credit	$96 million
Total	$1,295 million

The IFC B loan was syndicated to 57 financial institutions from Asia, Australia, Europe, the Middle East, and the United States. The parallel commercial loans were provided by the same financial institutions in proportion to their participation in the IFC B loan. Most of the participants were interested in providing larger contributions than they were asked to put into the project. The Thai revolving credit facility was offered by local Thai banks.

Financing of the project was successful because of the following factors:

- The credibility of the two project sponsors. After completion, all loans have limited recourse to sponsors' assets—essentially limited-margin support under certain narrowly defined circumstances. Caltex has 40 years of experience in Thailand's petroleum sector. PTT is well known for efficiency and solid financial management.

Figure 10.5 Financing Structure of the Star Refinery in Thailand

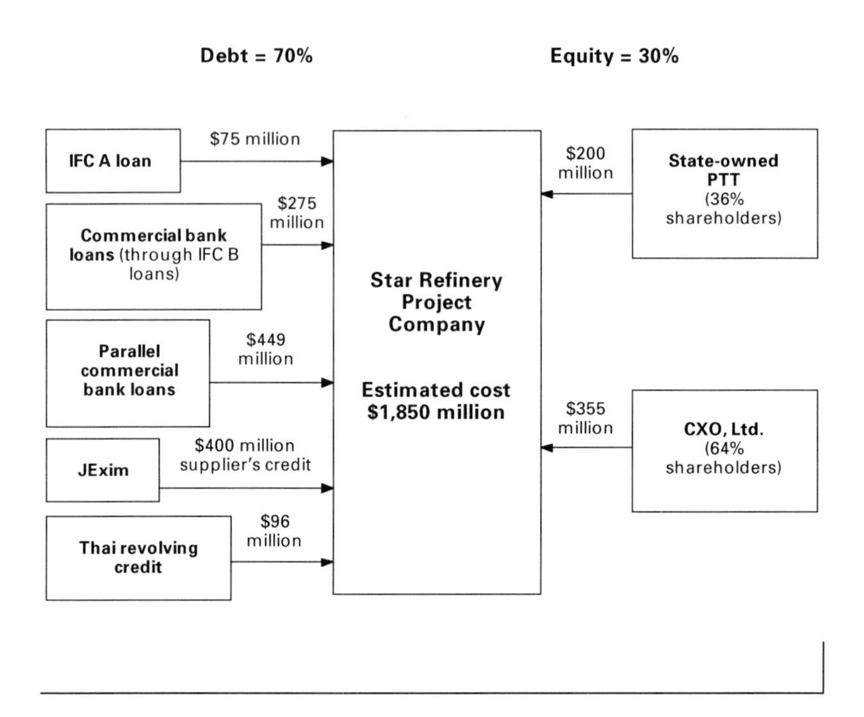

- The low market risk. Market demand is seen as very stable because of the solidity of Thailand's economy.
- The desirability of the business environment. In relation to market risk and political risk, government policies and other aspects of the country's business environment make Thailand a desirable location for foreign and private investment.

Financing of Natural Gas Infrastructure

The Bangkot Gas Transmission Project in Thailand. In most developing countries, gas infrastructure belongs to a state-owned public utility. Financing of gas infrastructure projects therefore has been based on the internal cash generation of the public utility, government budgetary contributions, and official borrowing. The Bangkot project, aimed at expanding Thailand's gas transmission capacity, is a typical example of

Figure 10.6
Financing Structure of the Bangkot Gas Project

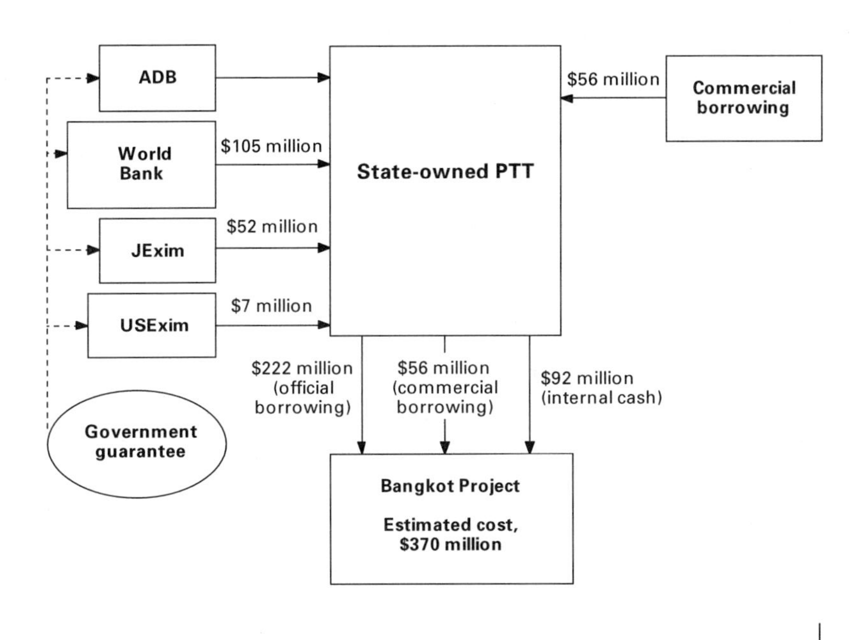

such projects. The project consisted of (1) a 32-inch, 175-km submarine pipeline from the Bangkot gas field to the existing Erawan production complex; (2) a 24-inch, 160-km pipeline from Erawan to a terminal at Khanom; and (3) associated facilities. The project was implemented in three years (1991-93) by PTT. The project cost was estimated at $370 million. The financing structure was as follows:

PTT's internal cash	$92 million
Asian Development Bank (ADB)	$58 million
World Bank	$105 million
JExim	$52 million
U.S. Export-Import Bank (USExim)	$7 million
Commercial borrowing	$56 million
Total	$370 million

"Commercial borrowing" above refers to PTT's borrowing from international commercial banks; bonds were issued on the local capital market. All the borrowing is on a corporate basis.

The Centragas Project in Colombia. Although many gas infrastructure projects in developing countries are still implemented by state-owned entities, there has been a decisive move toward utilization of private sector resources in such projects. The main impediments to private investment in this area are lack of well-established gas prices, substantial market risks, and unclear constraints imposed by regulations. However, private financing has been successful when project risks are shared in a straightforward manner. The Centragas project shows how such projects can work successfully.

Centragas is a limited-partnership company formed under Colombian law in 1994 to build, own, operate, and maintain a gas pipeline system for 15 years, after which the pipeline will be transferred to Ecopetrol, the state-owned oil company of Colombia. Centragas is wholly private, formed by Enron Development Corporation (a subsidiary of Enron).

The pipeline project, an integral component of Colombia's Gas Plan, includes an 18-inch, 575-km natural gas trunk line from Ballena on the northern coast to Barrancabermeja in the central region of Colombia and 21 lateral lines along the trunk line to connect with distribution networks, a dehydration facility, and two metering stations.

The total cost of the project was estimated at $217 million. The financing plan includes $45 million in equity contributed by the project sponsor and $172 million borrowed through private placement of bonds issued under SEC Rule 144A (see chapter 7). The bond issue carries an interest rate of 10.65 percent and a maturity of 16 years.

Centragas was able to achieve this simple financing structure primarily for the following reasons:

- Although the issuer of the bond is Centragas, the credibility of Enron as the main sponsor provides substantial comfort for potential lenders.
- Centragas signed a transportation service contract with Ecopetrol. According to this contract, Ecopetrol is obligated to pay a predetermined transportation fee to Centragas under most probable circumstances. Two monthly tariffs will be in effect: (1) a dollar-denominated

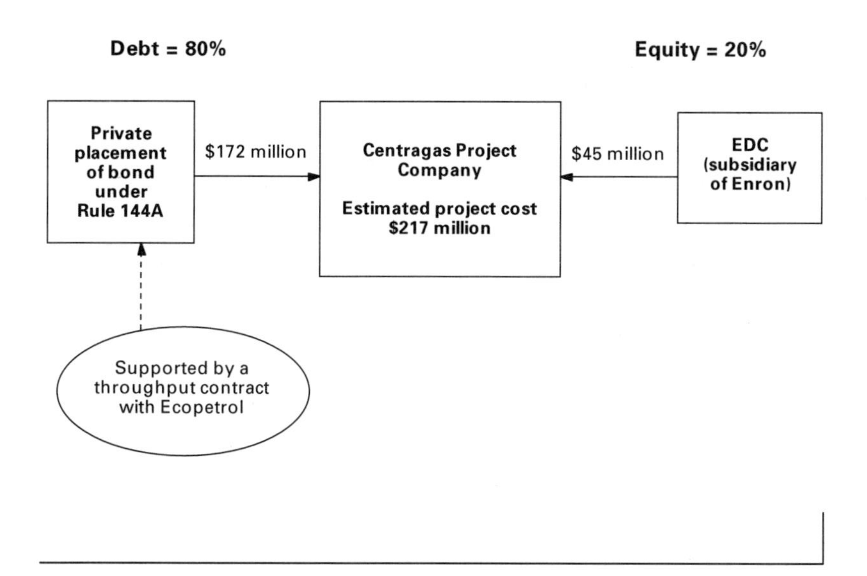

tariff designed to cover capital costs, debt service, certain taxes, and return on equity and (2) a dollar/peso-denominated tariff designed to cover the operation and maintenance (O&M) costs of the pipeline. Centragas will not be in the business of producing or selling gas.

- Centragas secured a fixed-price construction contract with a company (Tenco) experienced in building pipelines in Colombia.
- Enron International took responsibility for procurement of pipeline and technical advisory services.
- The bond issue was formally ranked at BBB- by Standard & Poor's Rating Group (S&P's).
- The Overseas Private Investment Corporation (OPIC) agreed to insure the project against political risks.

Financing Transnational Gas Projects

Financing transnational gas projects involving two or more developing countries has become quite challenging, primarily because of the large political and commercial risks involved. Many of these projects have remained at the preparation stage for more than a decade before being implemented. Financing becomes more feasible when reputable participants work closely with corresponding governments and international organizations to establish project viability, work out contractual arrangements, and bring investors and financiers together.

The Bolivian Gas Export Pipelines. In an interesting article, "Financing of Gas Pipelines in a Less Developed Country Two Decades Apart," Carlos Miranda, Bolivia's secretary of energy, compares the financing challenges of two gas export projects, a pipeline to Argentina built in the 1970s and a pipeline to Brazil expected to be built in the late 1990s.

The Bolivia-Argentina project was a 24-inch, 611-km pipeline (581 km were on the Bolivian side). The project cost of $58 million was financed by the World Bank; the Inter-American Development Bank (IDB); and a pension fund, the New York State Common Retirement Fund. Preparation and implementation of the project took only four years despite delays caused by Bolivia's nationalization of gas operations.

The Bolivia-Brazil project consists of a 28-inch, 1,800-km pipeline from Rio Grande in Bolivia to Campinhas in Brazil, and a 22-inch, 426-km pipeline extending to Curitiba. The cost of the project is estimated at $1.7 billion. The preparatory work for the project had, at this writing, lasted for more than 10 years. However, financing arrangements had not been finalized, although they were negotiated several times and presumably agreed upon.

Comparison of the above two projects leads to the conclusion that investors and financiers have become much more cautious about funding transnational pipelines in developing countries. The comparison also demonstrates that financing arrangements have generally become much more complex over the last two decades. The additional complexity takes its toll in time and cost.

The Algeria-Spain Gas Pipeline Project. This project involves a high-pressure gas pipeline system to transport gas from Algerian gas fields to the Spanish gas grid crossing Morocco and the Straits of Gibraltar. Capacity will initially be 9 billion cubic meters annually and could eventually reach up to 18 billion cubic meters with additional compression. The pipeline will be 1,385 km. In regard to ownership and the financing structure, the

Figure 10.8 Financing Structure of the Algeria-Spain Pipeline Project

A: Algerian Section

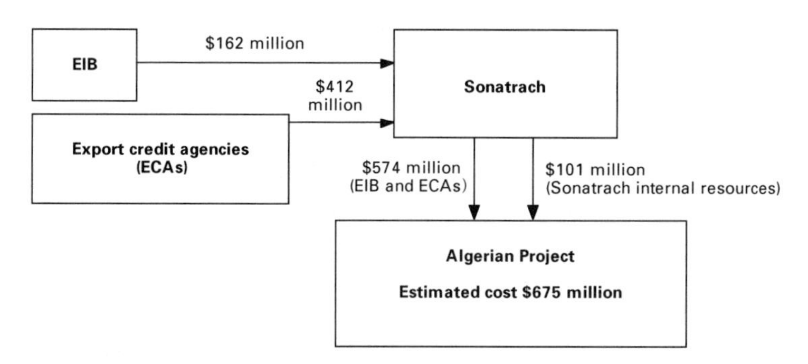

B: Moroccan Section

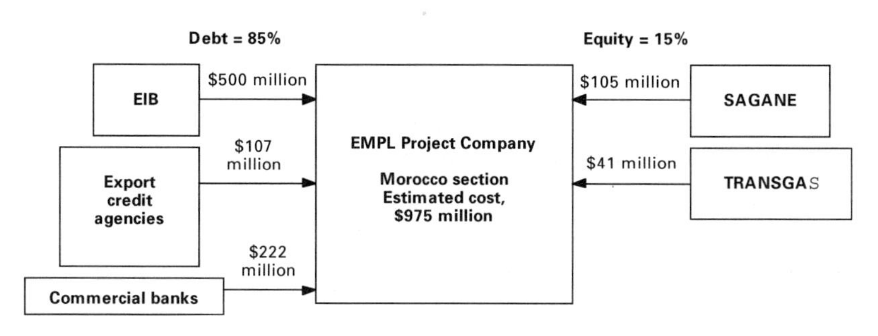

C: Spanish Section

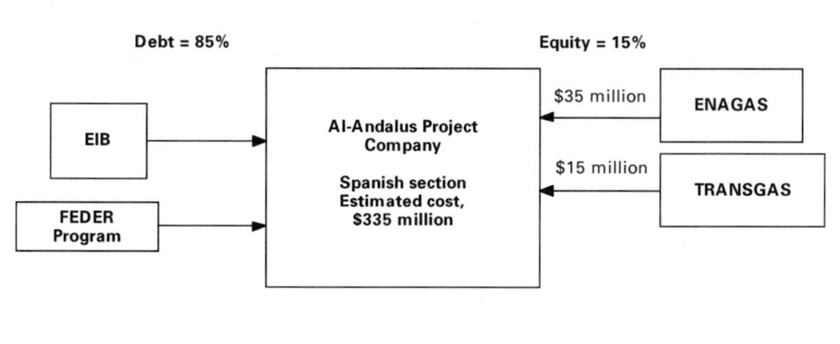

project is divided into three distinct sections: (1) the Algerian section, which will cost about $675 million; (2) the Moroccan section (from Morocco to the middle of the Straits of Gibraltar), which will cost about $975 million; and (3) the Spanish section (from the middle of the Straits of Gibraltar to Cordoba), which will cost $335 million.

The Algerian segment is owned by Sonatrach, the state oil and gas company, which finances 15 percent of the cost. The remaining 85 percent was financed by European Investment Bank (EIB; 24 percent) and several export credit agencies (USExim, COFACE, Hermes, and CESCE).

Maghreb-Europe Pipeline Limited (EMPL) holds exclusive right of use of the Moroccan section. Shareholders of the company are SAGANE of Spain (72 percent) and TRANSGAS of Portugal (28 percent). Concerning the financing structure, EMPL contributes 15 percent in the form of equity. The rest of the funding is provided by EIB (51 percent), export credit agencies (11 percent), and commercial banks (23 percent).

AL-ANDALUS, whose shareholders are ENAGAS (70 percent) and TRANSGAS (30 percent), has the exclusive right of use of the Spanish section. Regarding the financial sources, 15 percent of the funds are contributed by AL-ANDALUS as equity. The rest of the costs are financed by the European Union FEDER Program and the EIB.

The Nigeria LNG Project. The size, complexity, and risks of liquefied natural gas (LNG) projects reduce the options for financing. LNG projects ordinarily cost several billion dollars. Conventional financiers would not be able or willing to fund a large share of project costs unless for the expansion of a successfully operating system. Provided that the project is economic, attempts to mobilize funds for grassroots LNG projects are likely to succeed when

- The project sponsors and operators are large and very reputable international companies.
- The various components of the project (upstream, liquefaction plant, ships, and regasification facilities) are considered separately but are tightly linked in an "LNG chain."
- The project's size is adjusted to the financing possibilities.
- One reputable purchaser of gas, or several, would be willing to sign a take-or-pay contract with project sponsors.

- The project sponsors are able and willing to fund a large portion of project costs in the form of shareholders' equity.
- An overall commonality of interest exists among the decisionmakers involved in the various links of the LNG chain (expressed in substantial shareholding; strong, flexible contracts that are enforceable; or both) to ensure a fair and reasonable distribution of risks and benefits.

The Nigeria LNG project is an example of a difficult project in a difficult country, undertaken by reputable sponsors in a well-designed manner (after many attempts dating back to the late 1950s). Nigeria LNG Limited was incorporated under Nigerian law as a private, limited-liability company in 1989. The ownership structure (revised in 1993) consists of Nigerian National Petroleum Corporation (NNPC; 49 percent), Shell Gas BV (24 percent), Cleag Limited (an affiliate of Elf; 15 percent), Agip International BV (10 percent), and IFC (2 percent). The project is aimed at producing 5.7 million tonnes of LNG per year for transport from Nigeria to Italy, Spain, France, and the northeastern United States under long-term contracts that obligate the buyers either to take or pay for LNG made available by Nigeria LNG Limited. The take-or-pay contract is for 22.5 years. The buyers are reputable entities: ENEL of Italy, Enagas of Spain, Gas de France, and Distrigas Corporation of the United States.

Project facilities include a two-train liquefaction complex, a 218-km feedgas transmission system and related facilities. The upstream developments are carried out by separate joint ventures, although the upstream operators are also affiliates of Shell, Elf, and Agip. Shipping will be provided by Bonny Gas Transport Limited, a subsidiary of the project company. Construction will occur under a lump-sum turnkey contract with the TSKJ consortium, which comprises Technip, Snamprogetti, M.W. Kellogg, and JGC.

Funding of the project, estimated to cost $3.6 billion, was initially envisaged based on 47.6 percent equity and 53.4 percent debt financing. The equity funds were to be front loaded and spent before any debt was drawn down. Most of the debt, totaling $1.9 billion, was expected to be financed by export credit agencies ($1.5 billion); IFC was expected to provide $100 million in direct loans and arrange for $295 million in loans from a syndication of commercial banks. However, political circumstances in Nigeria made the export credit agencies hesitant to provide support. Project sponsors were convinced, eventually, of the need to fund almost all of the project

Figure 10.9
Financing Structure of the Nigeria LNG Project (Tentative Plan as of mid 1995)

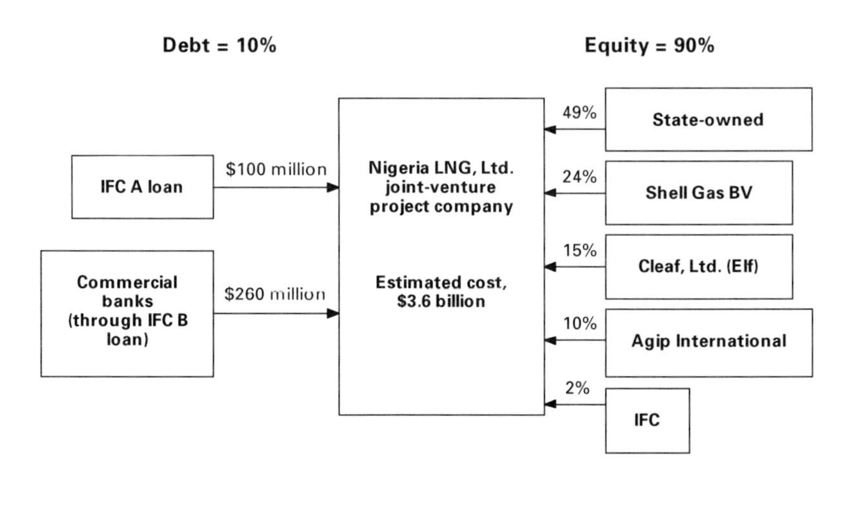

costs with their own resources, leaving open the possibly of refinancing project costs at a later stage. As a result, in mid-1995 they agreed to provide about 90 percent of the total project cost in the form of equity. The remaining 10 percent is expected to be financed by an IFC direct loan and an IFC syndicated loan from commercial banks.

Financing Power Generation Projects

Financing power generation projects in developing countries has for the past several decades been based on public budgets and public borrowing.

Since the mid-1980s, governments of developing countries have been receptive to private sector investment in building new generating capacity. At the initial stage, proposals were made in the form of build-operate-transfer (BOT), which meant that the private sector's capital and efficiency would be utilized to build the power plant, but after a number of years of operation the plant would be transferred to the public utility to be integrated into the rest of the country's generating capacity. As the ideas and proposals were further developed, it was recognized that the eventual transfer of the plant to the public utility was not a necessity. The proposals then changed to what became known as build-own-operate (BOO). This new arrangement is of course nothing but a straightforward private sector investment. The company is referred to as an independent power producer (IPP).

The Sarawak Power Project in Malaysia. A typical electricity project implemented by a state-owned utility is the Sarawak power project in Malaysia. The project consists of two 30-MW gas turbines and, associated with them, 185 km of double-circuit 275-kilovolt (kV) transmission line. The project was implemented in about three years (1990-93) by the Sarawak Electricity Supply Corporation (SESCO), which is responsible for power generation, transmission, and distribution in the state of Sarawak.

Like most other state-owned utilities, SESCO requests the approval of the government before proceeding with arrangements for financing and implementing a project. The government gives approval when the need for capacity expansion is clearly demonstrated, the choices of fuel and technology are justified, and the project is on the state's priority action list. Government approval also entitles SESCO to borrow funds from official sources based on the guarantee of the government.

The total cost of the Sarawak power project was estimated at $113 million. The financing structure was quite simple, comprising

SESCO's internal cash generation	$42 million
World Bank	$56 million
Export credit agencies	$15 million
Total	$113 million

The above financing structure required procurement of various components to be in conformity with government guidelines as well as with the guidelines of the external financiers. World Bank guidelines require that project components purchased with funds from World Bank loans must be procured through international competitive bidding. For components purchased with other funds, procurement is based on various considerations. For example, in the above project, conductors for transmission lines were reserved for local procurement to purchase domestically

Figure 10.10 Financing Structure of the Sarawak Power Generation Project

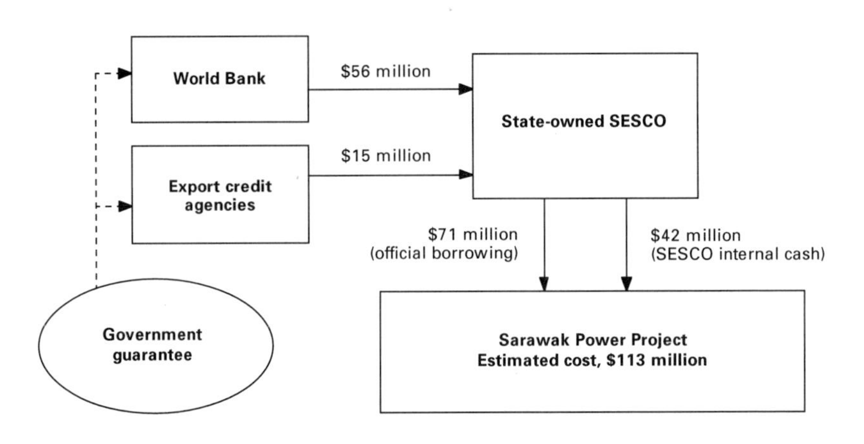

produced goods. Furthermore, switchgears, ancillary equipment, and transformers for substations were procured with export credits.

The Shajiao Power Project in China. The contract for the first successful BOT project in China, among the world's first, was signed in 1984 for construction of a coal-fired plant in Shajiao in Guangdong Province. It would be built by the Hong Kong-based Hopewell company and transferred to the Chinese government after 10 years of operation. The project was developed as a joint venture between the Shenzhen Special Economic Zone and a power development company. The terms of the joint venture were as follows:

- Hopewell was responsible for (1) arranging all foreign currency finance, (2) constructing the project, and (3) carrying out O&M for 10 years.
- Shenzhen was committed to (1) providing the land and operating staff, (2) arranging preferential tax treatment for Hopewell, (3) supplying all coal requirements at a fixed price, and (4) taking or paying for an agreed minimum amount of power at a predetermined price during the

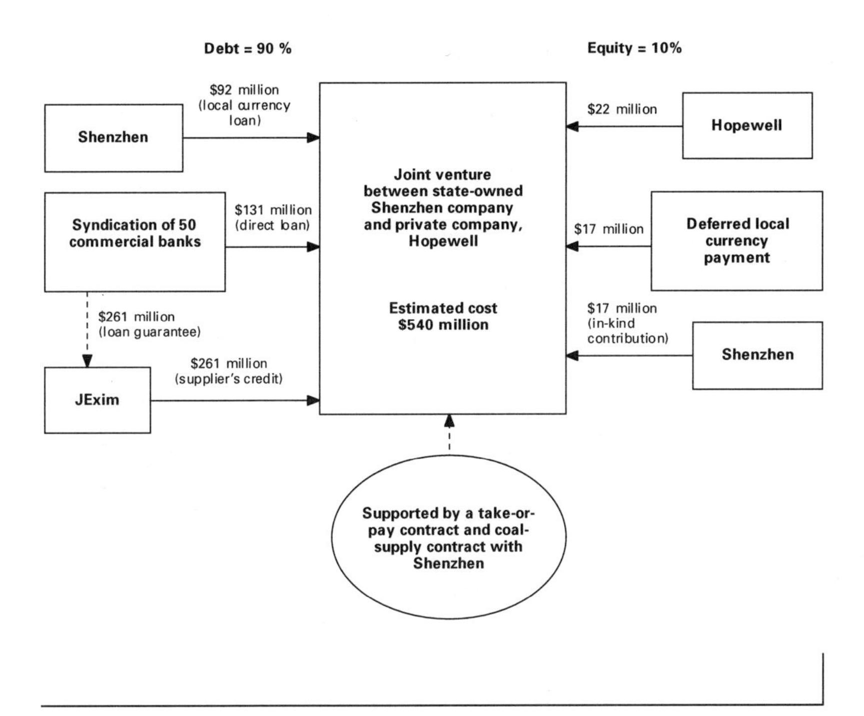

first 10 years of operation. The commitments of Shenzhen in the coal-supply and electricity take-or-pay contracts were guaranteed by a reputable financial institution, Guangdong International Trust and Investment Corporation (GITIC).

- Hopewell will retain 100 percent of the project revenue during the first 10 years of operation. It will cover costs appropriately. After 10 years of operation, full ownership and control will transfer from Hopewell to Shenzhen without compensation.
- Project revenue during the first 10 years of operation will consist of proceeds from the sale of electricity, 50 percent of which will be paid in Renminbi (local currency) and 50 percent in foreign exchange. The local currency portion will be used to pay for coal and other local items. The foreign currency portion will cover debt service and Hopewell's profits.

The total project cost was estimated at $540 million. Financing was highly leveraged; the shareholders' equity contribution was $56 million. Deferred local-currency payments of $17 million were included in this equity total. The debt was arranged as follows:

Local currency loan from Shenzhen equivalent to	$92 million
Supplier's credit from JExim	$261 million
Syndicated loans	$131 million
Total debt financing	$484 million

A syndicate of some 50 banks was arranged to provide the $131 million loan, as well as a project-risk guarantee to JExim for the $261.4 million supplier's credit.

Construction was based on a turnkey contract between Hopewell and an equipment-supply consortium arranged by Mitsui & Company including Toshiba Corporation, for turbines; Ishikawajima Harima Heavy Industries, for boilers; and Slipform Engineering (a Hopewell subsidiary), for civil engineering construction. The power plant was commissioned 28 months after the start of construction, 6 months ahead of schedule.

The Hub Power Project of Pakistan. Another well-known private power project is the Hub Power Project in Pakistan. Unlike the Shajiao project, the Hub project took a very long time to prepare. One important reason for delay was the complexity of financing arrangements. Because of these complexities, the project has become a classic case to be studied before any attempt to finance energy projects.

In the second half of 1980s the government of Pakistan opened the power sector to private investment. To encourage private investment in the sector, the government established the Private Sector Energy Development Fund (PSEDF), a government-owned facility to provide debt financing of up to 30 percent to private energy projects. The fund is supported by the World Bank, JExim, Commonwealth Development Corporation (CDC), France, Italy, and others (see chapter 6).

The first private power project in Pakistan was a 1,292 MW oil-fired power plant about 40 km outside of Karachi, built by the Hub Power Company, a special-purpose project company. Hub wanted to build the plant and sell power to the public utility, the Water and Power Development Authority (WAPDA). The total project cost was estimated at $1.8 billion, of which $1.7 billion was to be in foreign exchange and the remaining $100 million in local currency. The funding structure was as follows:

Equity investments	$372 million
Loan from the energy fund	$602 million
Loan from commercial banks	$695 million
Other local and foreign loans	$163 million
Total	$1,832 million

The complexity of the above funding scheme is primarily in the security structure, which involves a relatively large number of players. The following agreements were made between the project company and the relevant parties:

- A 30-year Implementation Agreement was made with the government, which granted the Hub company the sole right to develop the project.
- A 30-year Power Purchase Agreement was arranged with WAPDA, which secured the project's revenue.
- A 30-year Fuel Supply Agreement was concluded with the government-owned Pakistan State Oil Company.
- A fixed price turnkey construction contract was arranged with a consortium led by Mitsui & Company and including Ishikawajima Harima Heavy Industries Company, Ansaldo of Italy, and Campenon Bernard of France.
- A 12-year O&M agreement was made with a subsidiary of National Power (United Kingdom).
- Several other agreements were reached relating to local and offshore escrow accounts and foreign exchange risk insurance.

In addition, the World Bank, JExim, and a number of export credit agencies provided comprehensive guarantees to the syndicate of commercial banks for their corresponding lending to the project. The World Bank and JExim guarantees covered $240 million and $120 million, respectively. The remaining $335 million of commercial loan was guaranteed by export credit agency insurance. The World Bank and JExim guarantees backed the commitments made to the project by the government of Pakistan. The guarantees would be triggered if the government failed to comply with one or more of its obligations, as outlined in project contracts, which

Figure 10.12 Financing Structure of the Hub Power Project in Pakistan

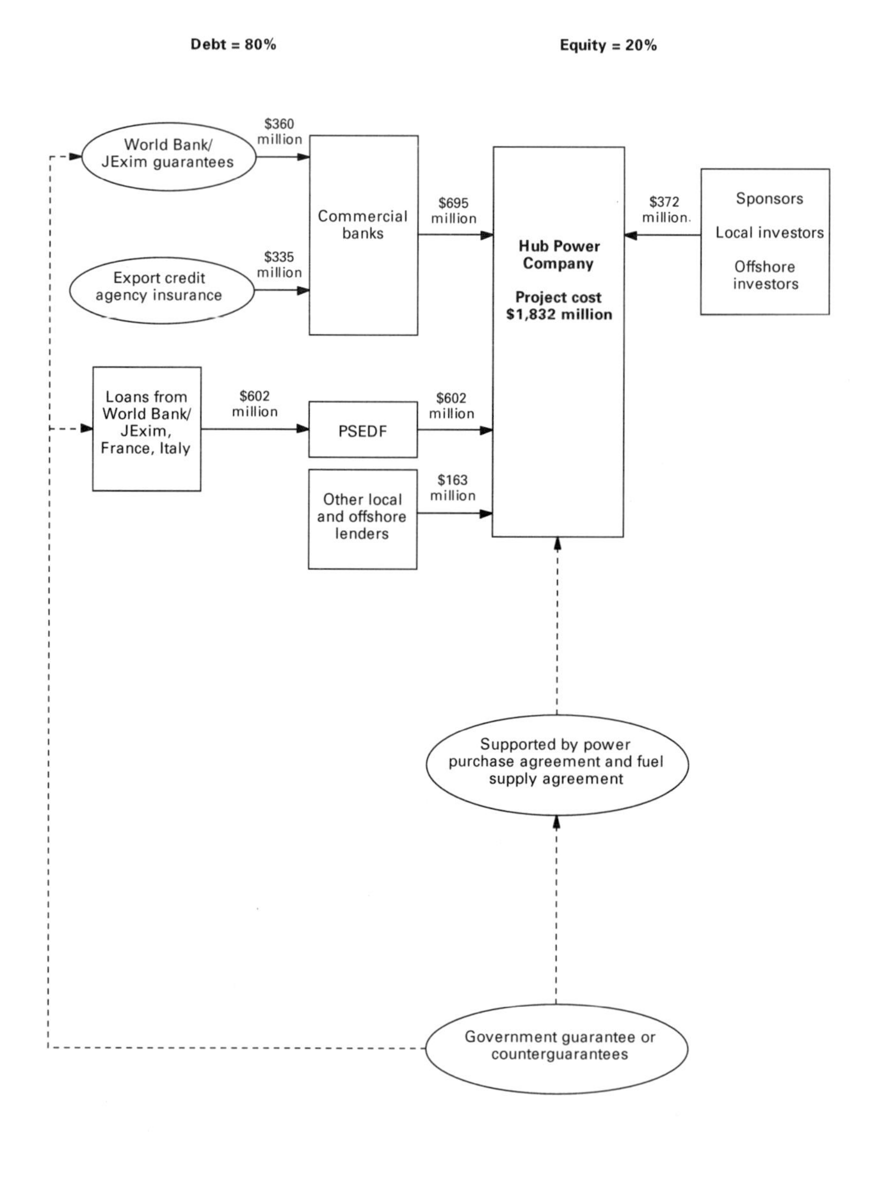

would constitute a default on the loans. These World Bank and JExim guarantees covered three categories of risk:

- Government guarantees of obligations of WAPDA and Pakistan State Oil Company regarding purchase of power and fuel supply, respectively.
- Government obligations included in the Implementation Agreement, including payments resulting from occurrence of certain events of *force majeure* (these can be political, such as war, or natural, such as a lightning strike affecting project equipment or facilities).
- Provision and transfer of foreign exchange through the Foreign Exchange Risk Insurance Scheme provided by the State Bank of Pakistan.

The Rockfort Power Project in Jamaica. After almost a decade of experience, the major players seem to have learned the most successful ways of dealing with hurdles involved in financing IPPs in developing countries. The Rockfort project is an example of an efficient undertaking in this area. The project is also noteworthy because of its extensive use of the bond market and tax-exempt financing to reduce the cost of equity and debt to the project.

The project consists of a 60-MW diesel power plant in Rockfort. The project cost is $144 million, which includes the base cost ($120 million) as well as standby facilities ($24 million).

Project sponsors include affiliates of Hydra-Co Enterprises, Incorporated (at the time, a subsidiary of a utility in New York State); U.S. Energy Corporation; Precursors Systems Incorporated; and International Energy Finance (IEF). This group of private power developers formed a special-purpose project company in Jamaica. The financing structure is based on 30 percent equity and 70 percent debt. The four sponsors committed $16 million in equity. The remaining equity was committed by the United Kingdom's CDC ($7 million); UtilCo Group, Incorporated (a subsidiary of a U.S. investor-owned gas and electric utility, $12 million); and the Energy Investors Fund (a private investment fund in the United States, $8 million). The debt is financed by CDC ($20 million) and by the bonds of the Caribbean Basin Projects Financing Authority (CARIFA), with a 5-year maturity, which are supported by a consortium of international banks through a $83 million letter of credit issued by the Deutsche Bank. Following project completion and at the maturity of the CARIFA bonds, term financing will be provided by the Jamaica Private Sector Energy Fund (PSEF, $81 million). PSEF is a government-owned lending facility financed by the World Bank and IDB, which each provide $40.5 million. Project sponsors also secured a $50 million guarantee from the Multilateral Investment Guarantee Agency (MIGA) to insure equity and debt against political risk.

Thus, the "apparent" financing structure is as follows:

Equity (30 percent)	
Hydra-Co	$16 million
UtilCo	$12 million
Energy Investors Fund	$8 million
CDC	$7 million
Equity Total	$43 million
Debt (70 percent)	
PSEF	$81 million
CDC	$20 million
Debt Total	$101 million
Total Financing	$144 million

In the above financing structure, except for CDC, none of the above parties put any funds, direct or apparent, into the project. Instead, they raised $122 million by issuing bonds, also referred to as "936 funds." The title of the fund relates to a section in the U.S. tax code that provides for a tax exemption on earnings of U.S. corporations in Puerto Rico, provided that these earnings are reinvested in Puerto Rico or certain countries (including Jamaica) in the Caribbean. Because of the reinvestment obligation, these funds carry low interest rates (below the London Interbank Offer Rate).

Sponsors of the Jamaica private power project raised $122 million through CARIFA, the Puerto Rican financial institution set up to channel investment funds from United States corporations to private sector projects in the Caribbean. CARIFA offered three bond issues, two totaling $43.2 million to fund the equity and one for $81 million to fund the debt. The latter was supported, as mentioned earlier, by a letter of credit from a commercial bank rated AAA. The equity portion of the issue has a maturity of two years and is indirectly backed by commitment from the equity investors through a letter of credit issued by Banco Santander. The debt portion of the issue has a maturity of five years and is indirectly backed by the PSEF (which is supported by the World Bank and IDB). The five-year bond issue was rated AAA, and the 2-year bond issue was rated AA-, by S&P's, based on the strength of the supporting letters of credit.

Figure 10.13 Financing Structure of the Rockfort Power Project in Jamaica

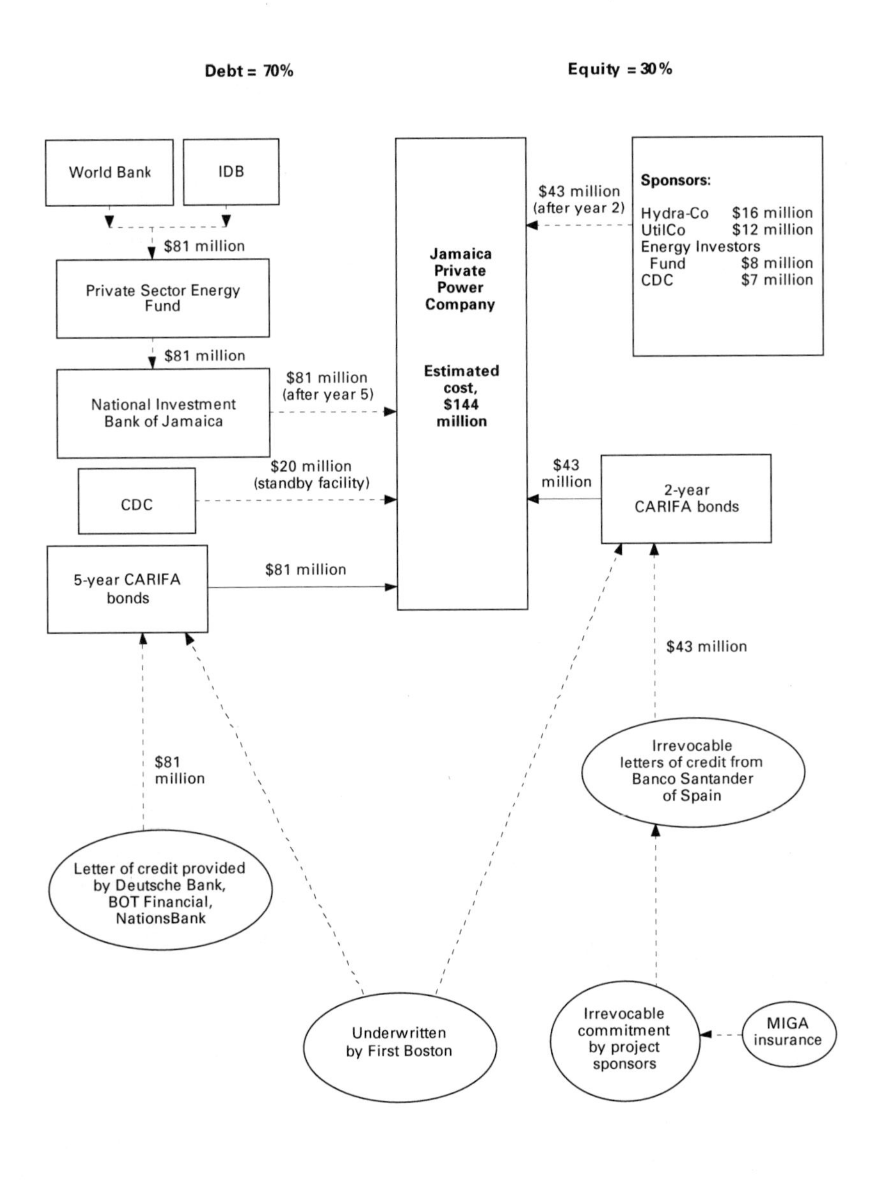

Figure 10.14 Financing Structure of the Distribution System Project in Thailand

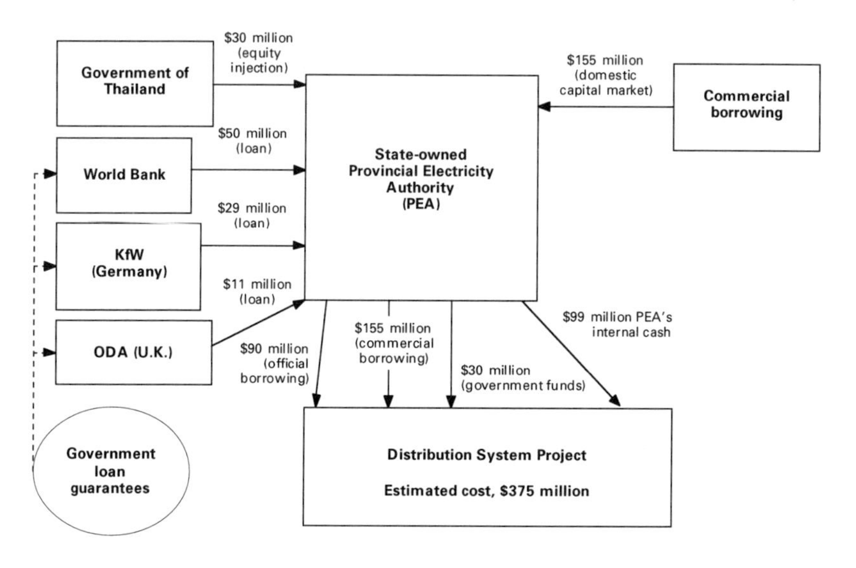

The successful mobilization of financing was attributable to the following factors:

- The securing of a 20-year power purchase agreement with the state-owned power utility. The agreement was further guaranteed by the government.
- The signing of an implementation agreement between the project company and the government, which set the general parameters for the company's investment in power generation.
- The signing of a 20-year fuel supply agreement with the government-owned Petrojam Refinery.

- The support of international commercial banks, which issued letters of credit based largely on the project sponsors' guarantee of repayment of equity bonds, and the World Bank and IDB funding agreements made through PSEF.
- The underwriting of both bond issues by First Boston.
- The guarantee by MIGA against political risks.

Power Transmission and Distribution

Power T&D projects are primarily implemented by state-owned utilities. Financing of these projects is arranged with governmental assistance. Government support is normally in the form of budgetary contribution or sponsorship for the utility's borrowing.

An example of a power T&D project is the effort to reintroduce the distribution system in Thailand, which is being implemented by the Provincial Electricity Authority (PEA). The PEA is a state enterprise responsible for distribution of electricity throughout Thailand, except in the metropolitan Bangkok area, which is serviced by another state company. The project is aimed at expanding the provincial 115-kV subtransmission system and extending 22-kV feeder networks to 1,000 unelectrified villages and 10,000 partially electrified villages. The project cost was estimated at $375 million.

The financing structure is as follows:

Internal cash	$100 million
World Bank	$50 million
KfW (Germany)	$29 million
Overseas Development Administration (ODA, U.K.)	$11 million
Commercial borrowing	$155 million
Equity injection by government	$30 million
Total	$375 million

The above financing structure is typical of T&D projects with some underlying notion of poverty reduction. The loans from KfW and ODA carry annual interest rates of 2 and 1.7 percent, respectively. The commercial borrowing is from domestic capital markets and carries an annual interest rate of about 10 percent.

11

ENVIRONMENTAL CONCERNS

Since the second half of the 1970s, growing environmental awareness has been accompanied by a more focused attention on the interactions between the natural environment and economic development. Although part of this attention has been translated into overall development policies, the main tangible strategy for minimizing damage to the environment has been to set specific environmental standards for construction and operation of commercial facilities, particularly energy producing and processing units. These standards are imposed not only on new plants but sometimes on plants already in operation. Indeed, application of stringent environmental standards to existing plants has caused much concern and in some cases tremendous financial loss.

Concerns about the environmental impacts of a project fall into three categories:

- Legal issues. Most developing countries have a legal requirement to assess the environmental impacts of energy projects and to ensure that standards are met during construction and operation of a plant.
- Possible financial damages. Experience has now demonstrated that returns from an investment, and even the investment itself, can be ruined if an energy plant does not adhere to environmental standards.
- Social responsibilities involving the local, regional, and global environments. These are much broader concerns raised by certain government agencies, bilateral and multilateral development agencies, and various environmental groups, as well as neighboring communities that may be affected.

The legal requirement to assess the environmental impact of any industrial project began in the United States in 1970, when the National Environmental Policy Act (NEPA) was signed into law by President Nixon. Today, most developing countries have passed similar laws, though details and enforcement vary considerably.

The task of assessing environmental impacts of a project has become increasingly complex as the framework of analysis has been broadened over time:

- When environmental assessment was conceived, it was regarded as an "add on" component to cost-benefit analysis, intended to incorporate the potential environmental damages into the cost of the project and to select the least-cost remedial actions.
- Soon, it became necessary to expand the framework to cover the direct and indirect impacts on climate and air quality, geology and soils, hydrology, and so on.
- It was further accepted that environmental assessments should be carried out by multidisciplinary teams able to examine social impacts as well.
- Associated with the above extension, it has become necessary to establish a link between environmental assessment and policy planning and to work with various interested and affected parties to ensure that all relevant views are taken into account.
- Project sponsors are often asked to go beyond the scope of the project and consider the environmental impacts of the entire fuel chain, or to examine how the remedial actions can be combined with other policies such as energy efficiency improvement strategies.
- Especially since the Rio Conference in 1992, examination of project impacts on the regional and global environment, although not typically a requirement of the host government, is demanded by some financiers.

Clearly, the above extensions to the examination of environmental aspects of projects have resulted in a complex, time-consuming, and sometimes unmanageable process. At the same time, however, the state of the art has progressed significantly, particularly with regard to communication with interested parties. Project sponsors are expected to be responsive to most if not all of these aspects at appropriate stages of project formulation, construction, and operation.

Overview 11.1
Environmental Issues in Project Analysis

- Financiers insist that project sponsors address relevant environmental impacts sufficiently because environmental problems can result in serious financial losses.
- Financiers have become increasingly committed to prevention of environmental damages that result from energy projects and set strict guidelines for review of the environmental impacts of such projects.
- Assessment of the environmental impacts of oil, gas, and power projects has become broader and more complex, covering many direct and indirect impacts, including impacts on climate and air quality, geology and soils, hydrology, social and economic structures, and sometimes regional and global environments.

THE OVERALL FRAMEWORK OF ENVIRONMENTAL ASSESSMENT

Environmental assessments (EAs) are required for all energy projects and by all financiers. Most financiers need a complete EA prior to project appraisal. The EA should be presented concisely and effectively to provide comfort that all important environmental risks have been identified and that a comprehensive plan has been formulated to mitigate and manage them. The World Bank Group has issued a set of guidelines for preparation of the EA. Most other financiers accept these guidelines. The World Bank guidelines for preparation of environmental assessment are summarized in Appendix C.

The essence of environmental assessment is the prediction of alternative future states of resources and environment that will result from various development paths, including no development at all. This is carried out by making the following assessments:

Measuring the environmental quality parameters before the project has begun. The quality of air and water supplies, fertility of soil, and nature and quality of habitat must be measured. These baseline data should also include natural changes that might occur in the absence of the proposed project.

Identifying significant impacts of the construction and operation of a plant on the environment. The World Energy Conservation Strategy suggests consideration of at least three criteria: first, the period and the geographic area over which the effect will be felt (including the number of people affected and the resources that would degrade); second, the urgency (that is, how quickly a natural system might deteriorate and how much time is available for its stabilization or enhancement); and third, the degree of irreversibility of damages to air, land, water, quality of life, plants, and animals.

Quantifying the environmental impacts. The objective is to evaluate the changes in the environment induced by a project that affect human health and welfare in either the short or long term. This would include measurement of eventual damages to human health, agricultural productivity, and other biophysical and socioeconomic changes likely to result from the project. Some of these damages may be measured on the basis of market prices; some are assessed in relation to international benchmarks and other proxies. However, some of the impacts remain nonquantifiable. These are analyzed qualitatively. Qualitative and quantitative assessments of environmental impacts of a project both incorporate the following aspects: First, assessment includes both positive and negative impacts of a project. Second, benefits and costs display a useful symmetry: a benefit forgone is a cost; a cost avoided is a benefit. Third, assessment is done on a "with- and without-project" basis, extended to alternative designs, sites, technologies, and modes of operation. Fourth, assessment includes costs and benefits of mitigation measures, including those reasonably proposed by concerned constituencies.

Overview 11.2 Environmental Assessments

- All financiers require an environmental assessment (EA) for energy projects.
- The EA examines (1) the baseline conditions, (2) the impacts of project construction and operation on these conditions, and (3) alternative development paths if the project is not constructed.
- Financiers have specific guidelines for preparation of EAs. The guidelines used by the World Bank Group are normally accepted by other financiers. These guidelines are summarized in Appendix C.

The remainder of this chapter explains briefly the specific environmental concerns that project sponsors should consider in preparing EAs for oil, gas, and power projects. Also summarized are the environmental standards usually used in EAs to demonstrate that environmental risks of the project are kept below harmful levels.

THE ENVIRONMENTAL IMPACTS OF ENERGY PROJECTS

Oil and Gas Development

Oil and gas production requires extensive on-site industrial activities for the life of the field. Construction of well sites, access roads, air fields, gathering and transport pipelines, and ancillary support facilities will result in extensive surface disturbance, construction traffic, noise and air emissions, and influx of construction crews. Continual noise and air emissions will be a part of equipment operation, as will discharge of treated formation waters and oil spills, which may affect soil and groundwater quality. Air pollution may also result from flaring of unwanted gases, sour-gas discharge (hydrogen sulfide), and burning of oil waste pits. Potential catastrophic accidents include well blowouts with uncontrolled oil and gas release and fires in refineries or gas-processing plants.

Offshore drilling and production facilities, vessel traffic, and coastal pipeline landings may interfere with fishing and with pleasure-boat use of the coastal area. Noise from overflights, near-shore drilling, production operations, port traffic, and processing plant operations will be distracting. Offshore and onshore facilities have visual impacts, as well. The initial construction force tends to be transient and is soon replaced by the operation staff, which is usually smaller and permanent.

Catastrophic oil spills from well blowouts, pipeline ruptures, or shipping accidents threaten marine mammals, sea and shore birds, and coastal areas. Control and cleanup of a major oil spill, well blowout, or fire—with emergency deployment of large crews, equipment, and supplies—creates a severe but temporary disruption of other activities in the coastal area. Residual effects from a spill can include oil-stained beaches, boats, and shore facilities and migration of certain marine species, at least temporarily.

Oil and Gas Pipelines

The major facilities associated with oil or gas pipelines include the pipeline itself; access or maintenance roads; the receiving, dispatch, and control station; digging stations; odorizing stations; and the compressor or pump stations. Because of internal friction and changes in elevation encountered along the line, booster stations are required at regular intervals for long-distance crude oil and product pipelines. For gas transmission lines, compression stations are installed at appropriate intervals along gas transmission lines to maintain pipeline pressures. Pipelines may transport unrefined oil or gas from a well-head to transfer or processing facilities. Refined oil and gas may be transported by pipeline to an end user, such as a petrochemical plant or a power plant, or to an export terminal.

Offshore, near-shore, and upland oil and gas pipelines have different environmental impacts. The magnitude of their impacts depends on the type and size of the pipeline installed; the significance depends on the degree to which natural and social resources are affected.

In the case of offshore pipelines, areas of concern include the following:

- The trenching and turbidity associated with pipelaying in offshore and near-shore areas may disrupt bottom-feeding organisms.
- Construction of the pipeline can temporarily resuspend bottom sediments. The subsequent resettlement of sediment may alter aquatic habitats and change the composition of species. The significance of these effects will depend on the type and importance of the aquatic organisms affected. For example, the alteration of seagrass beds or coral reefs, which are considered important feeding and breeding habitats for fish and other animals, may be greater than the alteration of deep offshore benthic habitats.
- If pipeline trenching occurs in near-shore and offshore areas where toxic chemicals have accumulated in the sediments, the laying of the pipeline can result in a resuspension of these toxic sediments and temporarily lower water quality immediately above the pipeline. Bioaccumulation of the toxic chemicals may occur in aquatic organisms (for example, fish and shellfish).
- In near-shore and offshore areas used for bottom fishing, pipelines can interfere with bottom trawling, resulting in loss or damage to fishing equipment as well as accidental ruptures to pipelines. Anchor dragging can also result in pipeline damage and oil spills.

For onshore pipelines, environmental damages can occur in the following areas:

- Installation of pipelines can promote erosion. In hilly areas, this can lead to instability and landslides. Runoff and sedimentation can lower water quality in rivers and streams during construction.

- Installation of pipelines and maintenance of roads can alter drainage patterns, blocking water flow and raising the water table on the upslope side of the pipelines, and can kill or reduce vegetation. If a pipeline cuts through a large forested area, the impact could be significant. Water supply to wetlands can be altered.
- Creation of rights-of-way (ROW) can introduce exotic plants able to replace native vegetation. In addition, pipeline installation can fragment the habitat of natural areas (for example, wildlands), resulting in the loss of species.
- In developed areas, oil and gas pipelines and substations can displace inhabitants or otherwise compromise land use. Some types of agricultural activities may be affected during construction.
- Above-ground pipelines can create barriers for humans and migratory wildlife. The significance of the effect depends on the length and location of the pipeline.
- Archeological sites are vulnerable to damage or loss during pipeline construction.
- Pipeline construction can temporarily interrupt traffic. This could be significant in developed areas if the pipeline crosses major transportation routes.
- Ruptures and leaks, as well as wastes generated at pump and transfer stations, can contaminate soils, surface water, and groundwater. The significance of this contamination depends on the type and size of the leak, type and volume of wastes generated, and the degree to which the natural resource is affected. Ruptures of oil pipelines crossing rivers and other water bodies or wetlands can result in significant environmental damage.
- Leakage or rupture of gas pipelines can cause explosions or fires.

Petroleum Refining

The environmental impacts of petroleum refining include gaseous emissions, wastewater discharges, solid wastes, noise, odor, and visual or aesthetic effects, and soil contamination. Atmospheric emissions include particulates, hydrocarbons, carbon monoxide, sulfur oxides, and nitrogen oxides. They emanate from various sources including the catalytic cracking units, sulfur-recovery processes, heaters, vents, flares, and product or raw material storage. Pump seals and valves can be sources of fugitive emissions. The combination of emissions can produce obnoxious odors.

Overview 11.3 Environmental Impacts of Oil and Gas Development

The environmental impacts of oil and gas development are reviewed with increasing scrutiny because of recent, well-publicized accidents. Major environmental impacts of oil and gas projects include

- Surface disturbances (traffic, noise, and so on).
- Air emissions from equipment operations, gas flaring, and burning of oil waste pits.
- Impact of treated formation waters and other wastes on water resources and contamination of soils, surface water, and ground water.
- Effects on aquatic resources and coastal activities from offshore pipeline laying.
- Effects on vegetation, land use, and wildlife from onshore pipeline construction.

Accidents can result in business disasters. Catastrophic oil spills from well blowout, pipeline rupture, or shipping accidents result in the rapid release of large quantities of oil into offshore waters. Cleanup of oil spills may be extremely difficult and expensive. Leakage or rupture of a gas pipeline can cause explosions or fires. In developed areas, such accidents pose significant human health risks.

The major pollutants present in petroleum refinery wastewater discharges are oil and grease, ammonia, phenolic compounds, sulfides, and organic acids, along with chromium and other metals. These pollutants may be expressed in terms of biochemical oxygen demand (BOD_5), chemical oxygen demand (COD), and total organic carbon (TOC). In addition, there is potential for serious surface water, soil, and groundwater contamination from leaks or spills of raw materials or products. Cooling water blowdown, flushing and cleaning water, stormwater runoff, and percolation from tank frames, pipe racks, product loading areas, and processing areas can also degrade surface water and groundwater.

Refineries generate large volumes of solid wastes; chief among them are catalytic fines from cracking units, coke fines, iron sulfides, filtering media, and various sludges (from tank cleaning, oil and water separators, and wastewater treatment systems).

Petroleum refining can be a noisy operation. Sources of noise include high-speed compressors, control valves, piping systems, turbines and motors, flares, air-cooled heat exchangers, fans, cooling towers, and vents. Typical noise levels range from 60 to 110 decibels (dB) at a distance of 1 meter from the source.

A major release or spill of raw materials, products, or wastes can be environmentally catastrophic, especially to marine or aquatic ecosystems. As in any operation involving hydrocarbons, fire and explosion are constant possibilities.

Thermal Power Projects

Thermal power projects include gas-, oil-, and coal-fired conventional plants as well as combined-cycle, gas turbine, and diesel power plants. The major components of thermoelectric projects include the power system (that is, power source, turbine, and generator) and associated facilities, which may include the cooling system, stack-gas cleaning equipment, fuel storage and handling areas, fuel delivery systems, solid waste storage areas, worker colonies, electrical substations, and transmission lines. Environmental harm can occur both during construction and operation of these plants. Construction impacts are caused primarily by the following site-preparation activities: cleaning; excavation; earth moving; dewatering, dredging, or impounding streams and other water bodies; establishing laydown areas; and developing borrow and fill areas. The large number of workers employed in constructing power plants can have significant sociocultural impacts on local communities.

The environmental impacts during plant operation include

- The impact of air emissions on local, regional, and global environments.
- The impact of cooling water and waste heat on aquatic plant and animal communities.
- The impact on the local community.

Air Emissions. Thermal plants are considered major air emission sources that can affect local and regional air quality. Sulfur dioxide (SO_2), oxides of nitrogen (NO_x), carbon monoxide (CO), carbon dioxide (CO_2), and particulates (that may contain trace metals) are emitted from the combustion of fuels by thermoelectric projects. The amounts of each depend on the type and size of facility, type and quality of fuel, and manner in which fuel is burned. Dispersion and ground-level concentrations of emissions are determined by a complex interaction among the physical characteristics of the plant stack, physical-chemical characteristics of emissions, meteorological conditions at or near the site when emissions travel from the stack to the ground-level receptor, topographical conditions of the plant site and surrounding areas, and the nature of the receptor (for example, people, crops, and native vegetation).

Emissions from thermoelectric projects can act as precursors of acid precipitation, particularly when the fuel is coal, which has a high sulfur content. Acid precipitation accelerates deterioration of buildings and monuments, radically alters aquatic ecosystems and damages vegetation. Combustion of fossil fuel in thermoelectric plants also generates CO_2 and NO_x; global warming has been attributed to increased levels of CO_2 and NO_x in the atmosphere.

Cooling Water and Waste Heat. Many power generating plants that use steam also have once-through cooling systems. If the high volume of water used by large plants with these cooling systems is taken from natural water bodies, such as rivers or bays, aquatic organisms may be entrapped in the cooling system. This can significantly reduce populations of fish and shellfish, some of which may be commercially important.

Discharges of heated water can elevate ambient water temperatures, which can radically alter existing aquatic plant and animal communities, favoring organisms that are suited to higher temperatures. These new communities are then vulnerable to the opposite effect—sharp reductions in ambient water temperature—when plants shut down because of failure or scheduled maintenance.

Use of evaporative cooling towers reduces the volume of water that must be withdrawn for cooling to that needed to offset evaporation. Towers eliminate thermal discharge but produce cooling-water blowdown, which must be discharged. In colder climates, a beneficial alternative is to reduce the temperature of cooling-water discharge by using waste heat to make hot water or steam for heating buildings or aquaculture ponds.

Either form of cooling entails some loss of water, reducing the volume available for drinking, irrigation, navigation, and other uses in water-short areas.

Impacts on the Community. One of the major impacts from power plants involves the influx of workers for building the plant. Several thousand workers may be required during the several years of construction of a large plant, and several hundred workers will be needed thereafter for its operation. There is potential for great stress where the host community is small. A "boomtown" effect, or condition of induced development, can result. This strains the existing community infrastructure—schools, police and fire protection, medical facilities, and so on. Similarly, the influx of workers will change local demographic patterns and disrupt local social and cultural values, as well as the living patterns of the residents. Another potential impact is the displacement of the local population because of land requirements for the plant site and associated facilities. Significant disruption of local traffic can occur from the construction and operation of a thermoelectric plant. In addition, large power plants can be visually obtrusive and noisy.

Hydropower Projects

Hydroelectric projects include dams, reservoirs, canals, penstocks, powerhouses, and switchyards for the generation of electricity. The dam and reservoir may be multipurpose; if watershed rainfall, stream-flow characteristics, and water and power usage patterns permit, hydroelectric reservoirs can also provide one or more of the following services: irrigation, flood control, water supply, recreation, fisheries, navigation, sediment control, ice-jam control, and glacial lake outburst control. However, these are competing uses for water stored behind dams, and each may imply a different diurnal or annual operating rule curve for the reservoir.

The principal sources of impacts in a hydroelectric project are the construction and operation of a dam and reservoir. Large dam projects cause major environmental changes over wide geographic areas.

The area of influence of a dam and its reservoir extends from the upper limits of the reservoir to as far downstream as the estuary, coastal, and offshore zones and includes the reservoir, dam, and river valley below the dam. Although direct environmental impacts are associated with the construction of the dam (for example, dust, erosion, borrow, and disposal problems), the greatest impacts result from the impoundment of water, which floods land to form the reservoir and alters the flow of water downstream. These effects have direct impacts on soils, vegetation, wildlife and wildlands, fisheries, climate, and human populations in the area.

The dam's indirect effects include those associated with the building, maintenance, and functioning of the dam (for example, access roads, construction camps, and power transmission lines) and the development of agricultural, industrial, or municipal activities made possible by the dam.

Environmental aspects of hydropower projects have become increasingly difficult to deal with. Indeed, some critics claim that the environmental costs of dams outweigh their benefits and that the construction of large dams is unjustifiable. Others contend that in some cases environmental costs can be avoided or reduced to an acceptable level by implementing cost-effective corrective measures. Although for some hydro projects, at least, the environmental impacts are undoubtedly manageable, project sponsors still must be prepared to go through a long and complex process of convincing financiers that the environmental risks are under control.

Power Transmission Projects

Electric power transmission lines affect the environment during construction, operation, and maintenance. Clearing of vegetation from sites and rights-of-way, along with construction of access roads, tower pads, and substations, are the primary sources of construction-related impacts. Operation and maintenance of the transmission line involve chemical or mechanical control of vegetation in the right-of-way and occasional line repair and maintenance. These, plus the physical presence of the line itself, can affect the environment.

The effects of short transmission lines can be localized; however, long transmission lines can have regional effects. In general, the environmental impacts on natural, social, and cultural resources increase with line length. The magnitude and significance of the impacts also increase as the voltage of the lines increases and requires larger supporting structures and rights-of-way. Operational impacts also increase with voltages. For example, electromagnetic field (EMF) effects are significantly greater for 1,000-kilovolt (kV) lines than for 69-kV lines.

On the positive side, power line rights-of-way, when properly managed, can be beneficial to wildlife. Cleared areas can provide feeding and nesting sites for

Overview 11.4 Environmental Impacts of Power Generation Projects

- Power generation projects are particularly subject to pronounced environmental concerns.
- Environmental concerns are raised about all types of power generation (coal, oil, gas, hydro, and nuclear); however, gas-based power plants are considered least harmful.
- Environmental impacts can occur during plant construction and operation.
- Construction impacts for thermal power projects are caused during the preparation of the site because of large numbers of workers employed to build the plant and the relocation of residents from the plant vicinity.
- Environmental impacts during plant operation include hazardous air emissions, effects of cooling water and waste heat on aquatic plant and animal communities, and socioeconomic impacts on local communities.
- The main impact of hydroelectric projects occurs during construction of the dam and reservoir and stems from the impoundment of water, flooding of land, and alteration of waterflow downstream, which irreversibly affect soils, vegetation, wildlife and wildlands, fisheries, climate, and human populations in the area.

birds and mammals. Power lines and structures can also serve as nesting sites and perches for many birds, especially raptors. Moreover, the increased habitat diversity resulting at the contact between the right-of-way and the existing vegetation is well-documented in biological literature as the "edge effect."

ENVIRONMENTAL STANDARDS

In preparing an EA for an energy project, sponsors must demonstrate that the environmental protection plan conforms with established standards. These standards refer to the limits and boundaries of environmental risks. For example, the maximum emission of sulfur dioxide from a power plant should stay below a standard to ensure that the ambient air quality does not fall below an acceptable level. The most frequently used standards relate to air and water quality. Many countries have two sets of air quality standards, one referring to ambient air and the other to allowable emissions. In the same way, standards of water quality refer to ambient water quality and allowable effluent levels. The ambient standards for air and water are set at levels below which no significant damage to the environment is likely. Emission levels for air and effluent levels for water represent the maximum allowable discharge from a point. These standards could be set at different levels depending on prevailing ambient conditions, location of the plant, and so on.

For most developing countries, environmental standards are of recent origin. Indeed, many countries do not have environmental standards. However, worldwide attention to environmental concerns has forced most financiers to require that a proposed energy project meet acceptable environmental standards. As a result, most developing countries are devising standards based on the experiences of industrialized countries and international agencies.

Environmental standards of industrialized countries vary substantially in coverage and application. In some countries, standards vary depending on the size, type, and location of the plant. Some countries have different standards for existing versus new plants. Some countries have different standards for various seasons of the year. Thus, no unique set of standards exists for developing countries. In addition, environmental standards of industrialized countries are often viewed as too stringent for use in developing countries.

Although there are no generally accepted international environmental standards, sets of standards are issued by the European Union (EU), World Health Organization, and other multilateral organizations. Financiers reviewing energy projects also use standards. The latter standards are most relevant to project preparation. Indeed, most EAs refer to these standards in order to establish environmental viability of a proposed project.

Environmental standards used by various financiers are often similar to the standards used by the World Bank Group. The World Bank does point out that its standards should be viewed as guidelines and judged in the context of plant location and existing environmental conditions. The standards show the maximum levels of hazardous air emissions and liquid effluents for energy projects. Because of the basic differences in technologies and processes, the standards are different for oil and gas development than for power generation projects. Some standards, however—for example, those for workplace air quality and noise level, and for handling, storage, and safety of materials—are almost the same for oil, gas, and power projects.

Oil and Gas Development

The environmental standards and guidelines for oil and gas development include recommendations for changing certain processes and specify limits for air emissions and liquid effluents.

Overview 11.5 Environmental Standards

- In preparing EAs for energy projects, sponsors need to demonstrate that environmental impacts remain within acceptable standards.
- Most developing countries are preparing standards based on the practices of industrialized countries and the guidelines used by international agencies.
- Although no generally accepted international environmental standards exist, several sets of standards—issued by the European Union, the World Health Organization, and multilateral development institutions—are available.
- Environmental standards used by financiers are often similar to standards issued by the World Bank Group. The standards are provided separately for oil and gas projects and for power projects. They describe (1) recommended practices and processes; (2) maximum levels of hazardous air emissions and liquid effluents; and (3) general requirements for workplace air quality, noise level, and so on.

The recommended process changes are as follows:

- Maximize use of freshwater gel-based mud systems.
- Eliminate use of invert muds.
- Recycle drilling mud decant water.
- Prevent degradation of sweet wells by sulfate-reducing bacteria by using biocides.
- Select less-toxic biocides, corrosion inhibitors, and other chemicals.
- Minimize gas flaring.
- Store crude oil in tanks equipped with secondary seals and vapor recovery systems.

- Remove hydrogen sulfide and mercaptans from gases before flaring.
- Use knockout drums on flares to prevent emissions of condensates.
- Regenerate spent amines and spent solvents or send them off site for recovery.
- Use low-nitrogen oxide (low-NO_x) burners in process heaters.
- Provide spill prevention and control measures (incorporation of bunds and hard surfacing for areas around storage tanks; use of pressure-relief valves; and installation of high-level alarms).
- Recover oil from process wastewaters.
- Segregate storm water from process water.

Other requirements are as follows:
- Monitors for hydrogen sulfide must be installed wherever this gas may accumulate. The monitors should be set to activate warning signals whenever detected concentration levels of H_2S exceed 7 mg/m^3 (5 ppm).
- An assessment of oil spill risks must be conducted, and an oil spill contingency plan must be developed and implemented.

Oil and Gas Pipelines

Environmental guidelines specific to oil pipelines include

- Positive pipe corrosion control measures.
- A program of periodic inspection and maintenance.
- Pressure sensors connected to alarms and automatic pump shutdown systems.
- A metering system that provides continuous input/output comparison for leak detection.
- Engineering design that provides adequate protection from likely external physical forces.
- Accurate and complete records of all inspections, leak incidents, unusual events, and safety measures taken.
- Minimization of disturbance to natural vegetation, soils, hydrological regimes, and topography.
- Positive measures to control population influx to remote areas via the increased access created by the pipeline right-of-way and to prevent associated secondary impacts (for example, encroachment on the lands or preserves of traditional indigenous populations; uncontrolled exploitation of natural resources).

Table 11.1 Air and Liquid Emission Parameters for Oil and Gas Projects

The air emission limits are as follows:

Parameter	Maximum value
Volatile organic compounds (VOCs; including benzene)	20 milligrams (mg) per normal cubic meter (Nm^3) (mg/Nm^3)
Hydrogen sulfide (H_2S)	5 mg/Nm^3
Nitrogen oxides (NO_x)	
Gas-fired	240 mg/Nm^3 (or 65 nanograms per joule [ng/J]) (for gaseous fuels)
Oil-fired	360 mg/Nm^3 (or 100 ng/J) (for liquid fuels)
Odor	Not offensive at the receptor end

The liquid effluent standards are as follows:

Parameter	Maximum value
pH	6 to 9
BOD_5	50 mg/l
COD	250 mg/l
Total suspended solids	50 mg/l
Oil and grease	10 mg/l
Phenol	0.5 mg/l
Total residual chlorine	1 mg/l
Cyanide (total)	0.2 mg/l
Arsenic	0.1 mg/l
Cadmium	0.1 mg/l
Chromium (hexavalent)	0.1 mg/l
Lead	0.1 mg/l
Mercury	0.001 mg/l
Total metals	10 mg/l
Temperature increase	less than 3°C[a]

[a] The effluent should result in a temperature increase of no more than 3 degrees Celsius at the edge of the zone where initial mixing and dilution take place. Where the zone is not defined, use 100 meters from the point of discharge.

Refineries

Emissions. Control of air emissions normally includes the capturing and recycling or burning emissions from vents, product transfer points, storage tanks, and other handling equipment. Boilers, heaters, other combustion devices, cokers, and catalytic units may require controls for particulate matter (PM). Catalytic cracking units should be provided with particulate-removal devices.

Parameter	Maximum value
Particulate matter (PM)	50 mg/Nm^3
Nitrogen oxides (NO_x)	360 mg/Nm^3
Sulfur oxides (SO_x)	500 mg/Nm^3
Hydrogen sulfide (H_2S)	15 mg/Nm^3

Effluents. Refinery wastewater often requires a combination of treatment methods to remove oil and other contaminants before discharge. Separation of different streams (such as stormwater) is essential to minimize treatment requirements. Oil is recovered from slops via separation techniques such as

Table 11.2 Emission Parameters for Refineries

Parameter	Maximum value
pH	6 to 9
BOD_5	30 mg/l
COD	150 mg/l
Total suspended solids	30 mg/l
Oil and grease	10 mg/l
Chromium (hexavalent)	0.1 mg/l
Chromium (total)	0.5 mg/l
Lead	0.1 mg/l
Phenol	0.5 mg/l
Benzene	0.05 mg/l
Benzo(a)pyrene	0.05 mg/l
Sulfide	1 mg/l
Temperature increase	less than or equal to 3° C[a]

[a] The effluent should result in a temperature increase of no more than 3 degrees Celsius at the edge of the zone where initial mixing and dilution take place. Where the zone is not defined, use 100 meters from the point of discharge.

centrifuging. A typical system may include gravity separation, dissolved air floatation, biodegradation, and clarification. A final polishing step using filtration or treatment with activated carbon or chemicals may also be required.

Power Projects

Emissions. The following guidelines apply to new fossil-fuel-fired thermal power plants or units of 50 megawatt electric (MWe) or larger and have been set at levels that can be achieved by adopting a variety of low-cost options or technologies, including the use of clean fuels. For example, dust controls capable of 98 to 99 percent removal efficiency, such as electrostatic precipitators (ESPs) or baghouses, should always be installed. Similarly, the use of low-NO_x burners, usually with other combustion modifications, should be standard practice. The range of options for the control of SO_x is greater because of large differences in the sulfur content of different fuels and in control costs. In general, for low-sulfur, high-calorific fuels (<1 percent S), specific controls may not be required, whereas coal cleaning (when applicable), sorbent injection, or fluidized-bed combustion may be adequate for medium-sulfur fuels (1 to 3 percent S). Flue-gas desulfurization (FGD) or other clean coal technologies should be considered for high-sulfur fuels (>3 percent S).

Particulates. For a coal-fired plant or unit, the recommended removal efficiencies from exhaust gases are 99 percent for all PM and 98 percent for PM_{10} (particulates of 10 microns or less in size). These removal efficiencies are to be achieved at least 95 percent of the time that the plant or unit is operating. For all plants or units, emissions should not exceed 50 mg/Nm^3 for particulates under full-load conditions.

Nitrogen oxides. For a coal-fired plant or unit, a reduction in NO_x emissions of 40 percent (relative to the case in which no NOx controls are installed) is recommended for 95 percent of the time that the plant or unit is operating. This should correspond to an emissions level of 650 mg/Nm^3 (230 ng/J). For an oil- or gas-fired plant or unit, the recommended reduction rates are 25 percent and 5 percent, respectively, corresponding to emissions levels of 360 (100 ng/J) and 240 mg/Nm^3 (65 ng/J).

Sulfur oxides. It is recommended that total SO_x emissions from the power plant or unit should be less than 0.20 tonnes (t) per day MWe of capacity for the first 1,000 MWe plus 0.01 t per day MWe for the incremental over 1,000 MWe. The concentration of SO_x in flue gases should not exceed 2,000 mg/Nm^3.

Effluents. The following standards for effluents apply to all types of thermal power plants:

Table 11.3 Effluent Parameters for All Types of Thermal Power Plants

Parameter	Maximum value
pH	6 to 9
Total suspended solids	50 mg/l
Oil and	0.5 mg/l
Chromium (hexavalent)	0.1 mg/l
Copper	0.5 mg/l
Iron	1.0 mg/l
Nickel	0.5 mg/l
Zinc	0.5 mg/l
Temperature increase	less than or equal to 3° C[a]

[a] The effluent should result in a temperature increase of no more than 3 degrees Celsius at the edge of the zone where initial mixing and dilution take place. Where the zone is not defined, use 100 meters from the point of discharge.

Other environmental requirements for power projects include the following:

- Formulations containing chromates should be avoided in water treatment processes.
- Transformers or equipment containing polychlorinated biphenyls (PCBs) or PCB-contaminated oil should not be installed, and existing equipment using these chemicals should be phased out and disposed of in a manner consistent with the requirements of the host country.
- Processes, equipment, and central cooling systems involving the use or potential release to the environment of chlorofluorocarbons (CFCs) and halons should not be installed, and these chemicals should be phased out of use in existing systems and processes and disposed of in a manner consistent with requirements of the host country.
- Storage and liquid impoundment areas for fuels, raw and in-process materials, solvents, wastes, and finished products should be designed with secondary containment (for example, dikes and berms) to prevent spills and contamination of soil, groundwater, and surface waters.

Health and safety measures recommended to prevent electrocution include the following:

- Strict procedures must be followed for de-energizing and checking electrical equipment before maintenance work.
- Strict safety procedures must be implemented, including constant supervision during maintenance work on energized equipment.
- Personnel must be trained on revival techniques for electrocution.

General Requirements for Oil, Gas, and Power Projects

Workplace Air Quality. The standards for workplace air quality are as follows:

- Periodic monitoring of workplace air quality should be conducted for air contaminants relevant to employees' tasks and the plant's operations.
- Ventilation, air contaminant control equipment, protective respiratory equipment, and air quality monitoring equipment should be well maintained.
- Protective respiratory equipment must be used by employees when the exposure levels for welding fumes, solvents, and other materials present in the workplace exceed local or internationally accepted standards or the following threshold limit values (TLVs):

Carbon monoxide	29 mg/m^3
Hydrogen sulfide	14 mg/m^3
Nitrogen dioxide	5 mg/m^3
Sulfur dioxide	5 mg/m^3

- Monitors should be installed that activate an audible alarm when toxic gas concentrations exceed half of the above TLVs.

Workplace Noise. The requirements for limiting workplace noise are as follows:

- Feasible administrative and engineering controls, including sound-insulated equipment and control rooms, should be employed to reduce the average noise level in normal work areas.
- Plant equipment should be well maintained to minimize noise levels.
- Personnel must use hearing protection when exposed to noise levels above 85 dB.

Work in Confined Space. The requirements for work in confined spaces are as follows:

- All confined spaces (for example, tanks, sumps, vessels, sewers, and excavations) must be tested for the presence of toxic, flammable, and explosive gases or vapor and for lack of oxygen before individuals enter or occupy these spaces.
- Adequate ventilation must be provided to these spaces before workers enter them and during the time they occupy these spaces.
- Personnel must use air-supplied respirators when working in confined spaces that may become contaminated or deficient in oxygen during the period of occupancy.
- Observers and assistants must be stationed outside of confined spaces to provide emergency assistance, if necessary, to personnel working inside these areas.

Hazardous Material Handling and Storage. The guidelines for handling and storing hazardous materials are as follows:

- All hazardous (reactive, flammable, radioactive, corrosive, and toxic) materials must be stored in clearly labeled containers or vessels.
- Storage and handling of hazardous materials must be in accordance with local regulations and appropriate to their hazard characteristics.
- Fire prevention systems and secondary containment should be provided for storage facilities, where necessary or required by regulation, to prevent fires or the release of hazardous materials to the environment.

General Health. The guidelines for general health are as follows:

- Sanitary facilities should be well equipped with supplies (for example, protective creams), and employees—particularly those exposed to dust, chemicals, or pathogens—should be encouraged to wash frequently.
- Ventilation systems should be provided to control work-area temperature and humidity.
- Personnel required to work in areas of high temperature or high humidity should be allowed to take frequent breaks away from these areas.
- Pre-employment and periodic medical examinations should be conducted for all personnel, and specific surveillance programs should be instituted for personnel potentially exposed to toxic or radioactive substances.

General Safety. The general safety precautions and measures are as follows:

- Shield guards or guard railings should be installed at all belts, pulleys, gears, and other moving parts.
- Elevated platforms and walkways, stairways, and ramps should be equipped with handrails, toeboards, and nonslip surfaces.
- Electrical equipment should be grounded and well insulated and in conformance with applicable codes.
- Personnel should use special footwear, masks, and clothing for work in ardous materials.
- For work near molten or high-temperature materials, employees should be provided with nonslip footwear, gloves, safety glasses, helmets, face protection, leggings, and other necessary protective equipment.
- Eye protection should be worn by personnel where there is a risk of flying chips or sparks or where intense light is generated.
- Personnel should wear protective clothing and goggles where corrosive materials are stored or processed.
- Emergency eyewash and showers should be installed in areas containing corrosive materials.
- A safety program should be established for construction and maintenance work.
- A fire-prevention and safety program should be implemented and should include regular drills.

Training. The guidelines for training are as follows:

- Employees should be trained on the hazards, precautions, and procedures for the safe storage, handling, and use of all potentially harmful materials relevant to each employee's task and work area.
- Training should incorporate information from the material safety data sheets (MSDSs) for potentially harmful materials.
- Personnel should be trained in environmental, health, and safety matters including accident prevention, safe lifting practices, the use of MSDSs, safe chemical-handling practices, and proper control and maintenance of equipment and facilities.
- Training should also include instructions on responding to an emergency, including the location and proper use of emergency equipment, use of personal protective equipment, procedures for raising an alarm and notifying emergency response teams, and proper response actions for each foreseeable emergency situation.

Appendix A: The Coordinates for Finance and Guarantee Agencies

MULTILATERAL COFINANCIERS

The World Bank
(IBRD)
1818 H Street, NW
Washington, DC 20433,USA
Tel: (202) 477-1234
Fax: (202) 477-6391

International Finance Corporation
(IFC)
1818 H Street, NW
Washington, DC 20433,USA
Tel: (202) 473-7711
Fax: (202) 676-0365

The African Development Bank
(AfDB)
01 B.P. 1387, Abidjan 01
Ivory Coast (Côte d'Ivoire)
Tel: (225) 20-44-44
Fax: (225) 21-77-53

The African Development Fund
(AfDF)
01 B.P. 1387, Abidjan 01
Ivory Coast (Côte d'Ivoire)
Tel: (225) 20-44-44
Fax: (225) 21-77-53

The Nigeria Trust Fund
(NTF)
01 B.P. 1387, Abidjan 01
Ivory Coast (Côte d'Ivoire)
Tel: (225) 20-44-44
Fax: (225) 21-77-53

Asian Development Bank (ADB)
6 ADB Avenue, Mandaluyong
Metro Manila
P.O. Box 789
0980 Manila, Philippines

Tel: (632) 4444 (within Philippines)
(632)711-3851 (International)
Fax: (632) 741-7961
(632) 632-6816

European Investment Bank (EIB)
100, Boulevard Konrad Adenauer
Luxembourge-Kirchberg
L-2950 Luxembourg

Tel: (352) 43-79-1
Fax: (352) 43-77-04

European Bank for Reconstruction and Development (EBRD)
One Exchange Square
London EC2 A2EH, United Kingdom

Tel: (44-171) 338-6000
Fax: (44-171) 338-6100

The Inter-American Development Bank (IDB)
1300 New York Avenue, NW
Washington, DC 20577, USA

Tel: (202) 623-1000
Fax: (202) 623-3096

Nordic Development Fund (NDF)
P.O. Box 185
FIN-00171 Helsinki, Finland

Tel: (358-0) 1800-451
Fax: (358-0) 6221-491

Nordic Investment Bank (NIB)
P.O. Box 249
Fabianinkatu 34
FIN-00171 Helsinki, Finland

Tel: (358-0) 180-01
Fax: (358-0) 622-1584

The OPEC Fund for International Development
P.O. Box 995
A1010 Vienna, Austria

Tel: (43-1) 515-640
Fax: (43-1) 513-9238

Arab Fund for Economic and Social Development
Abdlatif Y. Al-Hamad
Director General/Chairman of the Board of Directors
P.O. Box 21923
Safat 13080
Kuwait City, Kuwait
Tel: (965) 484-4500
Fax: (965) 481-5750
481-5760

The Islamic Development Bank (IsDB)
P.O. Box 5925
Jeddah 21432, Saudi Arabia
Tel: (966-2) 636-1400
Fax: (966-2) 636-6871

BADEA-Arab Bank for Economic Development in Africa
Sayed Abdel Rahman El-Mahdi Avenue
P.O. Box 2640
Khartoum, Sudan
Tel: (through Int'l Operator) 7736-46 or 7737-09 or 7704-98
Fax: (through Int'l Operator) 7706-00
Telex: 23098/22739/22248 BADEX
Via Satellite (INMARSAT)
Tel: 873-1665-105
Fax: 873-1665-106 BADEX
Telex: 583-1665-105

BILATERAL COFINANCIERS

United Arab Emirates
Abu Dhabi Fund for Arab Economic Development
P.O. Box 814
Saees Bin Ghobash Building
Abu Dhabi, United Arab Emirates
Tel: (97-12) 725-800
Fax: (97-12) 728-890

AUSTRALIA
Australian Agency for International Development (AusAID)
G.P.O. Box 887
Canberra ACT 2601, Australia
Tel: (61-6) 276-4000
Fax: (61-6) 276-4880

The Export Finance and Insurance Corporation (EFIC)
P.O. Box R65
Sydney, NSW 2000, Australia
Tel: (61-2) 390-21-1
(61-2) 390-21-11
Fax: (61-2) 251-38-51

AUSTRIA

Ministry of Foreign Affairs Directorate General for Development Cooperation (DGDC)
Ballhausplatz 9
A 1014 Vienna, Austria
Tel: (43-1) 531-150
Fax: (43-1) 531-153-619

Austrian Export Credit Agency (OEKB)
Am Hof 4
A 1010, Vienna, Austria
Tel: (43-1) 531-270
Fax: (43-1) 531-27-690

BELGIUM

Belgian Administration for Development Cooperation (BADC)
4 Rue de Trone
1050 Brussels, Belgium
Tel: (32-2) 500-6211
Fax: (32-2) 500-6585

Office National du Ducroire (OND)
Square de Meeus 40
B-1040 Brussels, Belgium
Tel: (32-2) 509-4211
Fax: (32-2) 513-5059

CANADA

Canadian International Development Agency (CIDA)
200 Promenade du Portage
Hull, Quebec
K1A OG6, Canada
Tel: (819) 997-7615
Fax: (819) 953-5348

Export Development Corporation (EDC)
151 O' Connor Street
Ottawa, Ontario
K1A 1K3, Canada
Tel: (613) 598-2500
Fax: (613) 237-2690

DENMARK

Danish International Development Assistance (DANIDA)
Asiatisk Plads 2
DK-1448 Copenhagen K, Denmark
Tel: (45-3) 392-0000
Fax: (45-3) 154-0533

Export Credit Council (EKR)
Gl. Kongevejbo
DK-1050 Frederiksberg
Copenhagen, Denmark
Tel: (45-1) 31-38-25
Fax: (45-1) 31-24-25

FINLAND

Department of International Development Cooperation (DIDC)
Ministry for Foreign Affairs
Katajanokanlaituri 3
FIN-00160 Helsinki, Finland
Tel: (358-0) 134-161
Fax: (358-0) 622-2576

FINFUND
P.O. Box 391
FIN-00121 Helsinki, Finland
Tel: (358-0) 348-434
Fax: (358-0) 348-433

The Finnish Export Credit Ltd. (FEC)
P.O. Box 123
FIN-00131 Helsinki, Finland
Tel: (358-0) 134-111
Fax: (358-0) 174-819

The Finnish Guarantee Board (VTL)
P.O. Box 187
FIN-00131 Helsinki, Finland
Tel: (358-0) 661-811
Fax: (358-0) 651-181

France

Caisse Française de Developpement (CFD) Cite du Retiro 35-37, rue Boissy d' Anglas 75379, Paris Cedex 08, France	Tel: (33-1) 40-06-31-31 Fax: (33-1) 40-06-36-61
Fonds d' Aide et de Cooperation (FAC) Ministere de la Cooperation 20, rue Monsieur 75700, Paris, France	Tel: (33-1) 47-83-10-10 Fax: (33-1) 43-06-97-40
Banque Française du Commerce Extérieur (BFCE) 21, Boulevard Haussmann 75009, Paris, France	Tel: (33-1) 48-00-48-00 Fax: (33-1) 48-00-46-74
Compagnie Française d' Assurance pour le Commerce Extérieur (COFACE) 12 Cours Michelet 92065, Paris La Defense 10, France	Tel: (33-1) 49-02-17-70 Fax: (33-1) 47-73-81-97

Germany

Bundesministerium für Wirtschaftliche Zusummenarbeit und Entwicklung (BMZ) Friedrich-Ebert-Alee 114-116 53113 Bonn, Germany	Tel: (49-228) 535-0 (operator) Fax: (49-228) 535-3500
Kreditanstalt für Wiederaufbau (KfW) Palmengartenstrasse 5-9 60325 Frankfurt, Germany	Tel: (49-69) 7431-0 (operator) Fax: (49-69) 7431-29-44
Gesellschaft für Technische Zusammenarbeit (GTZ) Dag Hammarskjold-Weg 1 65760 Eschborn, Germany	Tel: (49-6196) 790 Fax: (49-6196) 79-11-15

Deutsche Finanzierungsgesell-schaft für Beteiligungen in Entwichlungs-laendern GmBH (DEG)
Belvedere-Strasse
500 Koln 41, Germany
Tel: (49-211) 498-6402
Fax: (49-211) 499-6290

Hermes Kreditversicherungs AG (Hermes)
Friedensallee 254
2000 Hamburg 50, Germany
Tel: (49-40) 887-0
Fax: (49-40) 887-9175

ITALY

Ministero degli Affari Esteri Direzione Generale per la Cooperazione allo Sviluppo (DGCS)
Piazzale della Farnesina, 1
00194 Rome, Italy
Tel: (39-6) 323-5885
Fax: (39-6) 323-5922

Istituto Centrale per il Credito a Medio Termine (MEDIOCREDITO CENTRALE)
Via Piemonte 51A
00194 Rome, Italy
Tel: (39-6) 47-911
Fax: (39-6) 323-5885

Sezione Speciale per Assicurazione del Credito all' Esportazione (SACE)
Piazza Poli 37
00100 Rome, Italy
Tel: (39-6) 673-6323
Fax: (39-6) 673-6225

JAPAN

The Export-Import Bank of Japan (JExim)
1-4-1 Ohtemachi, Chiyoda-Ku
Tokyo 100, Japan
Tel: (81-3) 3-287-1221

Japan International Cooperation Agency (JICA)
2-1-1 NishishinjukuShinjuku-Ku
Tokyo 163-04, Japan
Tel: (81-3) 3-346-5311
Fax: (81-3) 3-346-5032

The Overseas Economic Cooperation Fund (OECF)
1-4-1 Ohtemachi
Chiyoda-Ku
Tokyo 100, Japan
Tel: (81-3) 3-215-1311

KOREA, REPUBLIC OF

Export-Import Bank of Korea (KExim)
Attn: Economic Development Cooperation Fund Department
16-1, Yoido-dong, Yongdungpo-ku
Seoul, Korea
Tel: (82-2) 784-1021
Fax: (82-2) 769-1267

KUWAIT

Kuwait Fund for Arab Economic Development
P.O. Box 2921
Safat 13030
Safat, Kuwait
Tel: (965) 246-88-00
Fax: (965) 241-90-90

THE NETHERLANDS

Ministry of Foreign Affairs
DGIS, Bezuidenhoutseweg 67
P.O. Box 20061
2500EB, The Hague, The Netherlands
Tel: (3170) 3486-486
Fax: (3170) 3484-848

The Nederlandsche Credietverzekering Maatschappij N.V. (NCM)
Keizersgracht 271-285
1016 ED Amsterdam, The Netherlands
Tel: (31-20) 553-91-11
Fax: (31-20) 553-28-11

NORWAY

Norwegian Agency for Development Cooperation (NORAD)
Tollbugt 3
Oslo 1, Norway
Tel: (47-22) 33-55-60
Fax: (47-22) 31-44-01

A/S Eksportfinans
Dronning Maudsgt 15
BP 1601 Vika
N-250 Oslo 2, Norway

Tel: (47-22) 83-70-70
Fax: (47-22) 83-24-45

Guarantee Institute for Export Credits (GIEK)
Dronning Maudsgt 15
P.B. 1763 Vika
N-250 Oslo 2, Norway

Tel: (47-22) 83-70-70
Fax: (47-22) 83-24-45

PORTUGAL

Institute for Portuguese Cooperation (INCOOP)
Av. da Liberdade, 192, 2º
1200 Lisboa, Portugal

Tel: (351-1) 356-20-31
Fax: (351-1) 52-78-97

Fund for Economic Cooperation (FCE)
Av. da Liberade
258, 5º
1200 Liboa, Portugal

Tel: (351-1) 352-0607
Fax: (351-1) 315-4583

SAUDI ARABIA

Saudi Fund for Development
P.O. Box 1887
11441 Riyadh, Saudi Arabia

Tel: (966-1) 464-0292
Fax: (966-1) 464-7450

SPAIN

Agency for International Cooperation (ACI)
Ministry of Foreign Affairs
Reyes Catolicos, 4
28040 Madrid, Spain

Tel: (34-1) 583-8100
Fax: (34-1) 350-7988

Export Credit Insurance Company (CESCE)
Velasquez, 74
28001 Madrid, Spain

Tel: (34-1) 577-6066
Fax: (34-1) 576-5140

Institute for External Trade (ICEX)
Paseo de la Castellana 162
28046 Madrid, Spain
Tel: (34-1) 503-5737
Fax: (34-1) 458-2196

SWEDEN

Swedish International Development Authority (SIDA)
Sveavägen 20
S-105 25 Stockholm, Sweden
Tel: (46-8) 698-5000
Fax: (46-8) 673-2141

The Swedish Export Credits Guarantee Board (EKN)
P.O. Box 3064
S-103 61 Stockholm, Sweden
Tel: (46-8) 701-0000
Fax: (46-8) 411-8149

SWITZERLAND

Directorate for Development Cooperation and Humanitarian Aid (DEH)
Eigerstrasse 73
CH-3003 Berne, Switzerland
Tel: (41-31) 322-3475
Fax: (41-31)371-4767

UNITED KINGDOM

The Overseas Development Administration (ODA)
94 Victoria Street
London SW 1E 5JL, United Kingdom
Tel: (44-171) 917-7000
Fax: (44-171) 917-0019

The Crown Agents (CA)
1815 H St., NW
Washington, DC 20006, USA
Tel: (202) 822-8052
Fax: (202) 822-8064

The Commonwealth Development Corporation (CDC)
One Bessborough Gardens
London SW1V 2JQ, United Kingdom
Tel: (44-171) 828-4488
Fax: (44-171) 828-6505

Export Credit Guarantee Department (ECGD)
P.O. Box 2200
2 Exchange Tower
Harbour Exchange Square
London E14 9GS, United Kingdom

Tel: (44-171) 512-7000
Fax: (44-171) 512-7649

Department of Trade & Industry (DTI)
Projects & Export Policy Division
234 Victoria Street
London SW1E 6RB, United Kingdom

Tel: (44-171) 215-6210 or 6735
Fax: (44-171) 215-6535

UNITED STATES

U.S. Agency for International Development (USAID)
320 21st Street, NW
Washington, DC 20523, USA

Tel: (202) 663-2660
Fax: (202) 663-2670
663-2677

Export-Import Bank (USExim)
811 Vermont Avenue, NW
Washington, DC 20571, USA

Tel: (202) 565-3946
Fax: (202) 565-3380

Overseas Private Investment Corporation (OPIC)
1100 New York Avenue, NW
Washington, DC 20527, USA

Tel: (202) 336-8799
Fax: (202) 408-9859

U.S. Trade and Development Agency (TDA)
1621 North Kent Street, 3rd Floor
Rosslyn, Virginia 22209, USA

Tel: (703) 875-4357

Appendix B: Sources of Information for Preparation of Energy Projects

Information for preparation of energy projects is available from a variety of sources, many of which have become public only in recent years. Although some sources may duplicate items covered by others, all of the following sources should be consulted to arrive at a comprehensive and reliable picture of a developing country's business environment. For ease of use, the following sources are grouped into three areas: (1) country political and economic conditions, (2) country energy demand and supply, and (3) international energy outlook.

COUNTRY POLITICAL AND ECONOMIC CONDITIONS

The major sources of information are (1) the World Bank and regional development banks, (2) the International Monetary Fund (IMF), and (3) the Economist Intelligence Unit (EIU; located in London).

The World Bank and regional development banks publish annual statistics regarding national income account, balance of payments, external debt, inflation, and so on, for all member countries.

Some useful publications are:

- *World Tables*, published annually by the World Bank, provides time-series statistics on countries' macroeconomic variables, including resource flow, sectoral breakdown of GDP, governmental budget and expenditures, foreign trade balance of payment, and external debt.
- *Trends in Developing Economies*, published annually by the World Bank, includes a two or three page essay on each country's economy, recent developments, and prospects for medium-term growth. The volume includes data on GDP, government finance, balance of payments, and external debts.

- *Asian Development Outlook*, published annually by the Asian Development Bank (ADB), includes short essays on each Asian country's recent trends in growth, investments, government budget, external trade, as well as a two-year forecast of economic growth.
- *World Debt Tables*, published annually by the World Bank, includes time-series data on national debt as well as public and private flow of funds into developing countries.
- Economic reports by the World Bank and regional banks on individual countries. These reports are not published at regular intervals but are available for most developing countries. In 1994, the World Bank established a Public Information Center, which provides these reports to the public at a nominal cost.

The IMF publishes documents on each of its member countries' monetary aspects. Some of these documents contain results of specific research conducted on ad hoc bases. The following publications are published regularly and provide information regarding business environments of developing countries:

- *International Financial Statistics*, published annually, includes country time-series data on exchange rates, holdings of foreign exchange, gold and foreign assets, money supply, interest rates, government finance, national accounts and external trade.
- *International Capital Markets*, published annually, contains a review of several aspects of capital markets which includes recent developments in private market financing for developing countries.

The EIU publishes information about political and economic conditions of most developing countries. These two types of reports are published regularly:

- Country profiles, published annually, include detailed information on a specific country's political background, economy, natural resources, infrastructure and external trade.
- Country reports, published on a quarterly basis, include analysis of recent political, economic and sectoral trends. They also contain a short-term perspective on future political and economic conditions.

COUNTRY ENERGY DEMAND AND SUPPLY

Information regarding energy demand and supply in developing countries is published by the United Nations, the International Energy Agency

(IEA), the World Bank and other multilateral agencies and industry sources (such as British Petroleum, Cedigas, and the Oil and Gas Journal). UN publications include:

- *Energy Statistics Yearbook*, includes time-series on production, consumption and trade for all fuels in all countries. The data covers international movements of coal, crude oil, petroleum products, liquefied natural gas and uranium.
- *Energy Balances and Electricity Profiles*, published annually, contains country-by-country energy balances, energy conversion and electricity production.

The IEA conducts analyses of world energy demand and supply. A major part of this analysis concentrates on Organisation for Economic Co-operation and Development (OECD) countries. However, IEA also publishes data on energy flows in non-OECD countries. Relevant publications include:

- *Energy Statistics and Balances of Non-OECD Countries*, published annually, includes time-series on energy production, consumption, trade, as well as energy balance tables for most developing countries.
- *Energy in Developing Countries*, published as available, includes analyses of energy consumption and policies.

The World Bank and other multilateral agencies publish numerous reports on energy sector conditions in developing countries. However, these reports are not published at regular intervals. Publications that are especially helpful for preparation of energy projects include:

- Staff Appraisal Reports (SARs), prepared by the World Bank and regional development banks for each loan made to a country. The reports on oil, gas, and power loans contain analysis of the energy sector.
- Energy Sector Studies, conducted occasionally by multilateral institutions, include studies of the oil, gas, and power sectors. These studies provide detailed analysis of energy demand and supply, structural and policy issues, and forthcoming developments.
- Energy Sector Management Assistance Program (ESMAP) reports. ESMAP is managed by the World Bank and carries out strategic studies of energy issues and options in developing countries.

There are numerous industry sources for energy information. However, many of these sources are aimed at providing information about short-term perspectives and, therefore, are not particularly useful for preparation of investment projects. The following sources are useful for long-term statistics:

- *Natural Gas in the World*, an annual survey published by Cedigas, contains time-series data on volumes of gas production, flaring, venting, reinjection, export and import. It also contains discussions of trends in gas use and prices in major gas consuming countries.
- *International Petroleum Encyclopedia*, published annually by PennWell Publishing includes extensive maps listing locations of oil and gas reserves, refineries, terminals and transmission pipelines. It also contains reports on recent developments in upstream and downstream activities.
- *Statistical Review of World Energy and Statistical Review of World Gas*, published by British Petroleum, contain crisp, concise presentations of statistical trends in production, consumption, and international trade of coal, oil and natural gas.
- Several journals including the *Oil and Gas Journal, Power Engineering, Independent Energy, Independent Power*, and the *Petroleum Economist* provide coverage of energy issues and prospects in developing countries.

INTERNATIONAL ENERGY OUTLOOK

There are a number of institutions, including most major oil companies, that analyze the world energy outlook. The following two sources provide long-term forecasts on a regular basis and make their results available to the public:

- *World Energy Outlook*, published annually by the IEA, contains forecasts of world energy demand and supply under various assumptions. Worldwide and regional forecasts are presented.
- *International Energy Outlook*, published annually by the Energy Information Agency of the U.S. Department of Energy, contains forecasts of world energy demand and supply by type of fuel (coal, petroleum, gas). It also includes an analysis of the impact of energy prices and technology on the development and growth of the various types of energy. Worldwide and regional forecasts are presented.

Appendix C: Guidelines for Preparation of Environmental Assessments

Most financiers have published guidelines for preparation of environmental assessments (EAs). These guidelines are similar to those issued by the World Bank Group. This annex contains a summary of the World Bank guidelines. The details are published in a three-volume Environmental Assessment Sourcebook (1995).

An EA should include the following sections:

- Executive summary.
- Introduction.
- Legislative, regulatory, and policy considerations.
- Description of the proposed project.
- Description of the baseline environment.
- Potential impacts of the proposed project.
- Environmental management and mitigation plan.
- Monitoring plan.
- Coordination with government agencies, affected communities, and nongovernmental organizations (NGOs).
- References.

Executive Summary

The executive summary must be a concise description of the proposed project, environmental setting, significant findings, and recommended actions to mitigate significant impacts and monitor environmental performance. The executive summary should be no more than 20 pages of text, with one figure or map showing the project's location.

Introduction

This should contain the following items:

- Brief description of general location and type of activity.
- Identification of project sponsors.
- Brief history of the project, if project represents expansion, modernization, or restart of previous activity at the proposed site.

Legislative, Regulatory, and Policy Considerations

All applicable regulations and standards governing liquid effluents, air emissions, solid waste management, environmental quality, and health and safety must be identified and outlined, with numerical standards specified in summary tables. These include national and local regulations and standards as well as the guidelines and requirements of any financial institutions to which the sponsors apply for financing. The project's consistency with relevant World Bank policies on forestry, wildlands, and resettlement must also be described where appropriate.

Description of the Proposed Project

The EA must provide a description of the project, preferably using maps (at appropriate scale), figures, and tables where appropriate, and including the following information:

- Site or sites, including any installations or investments of the project sponsor that are directly linked to the proposed project even though outside of the main project site.
- Provision of services from offsite (for example, energy, water, or transportation).
- Process diagram showing the steps in the process, material flow and balances, and sources of liquid effluents, air emissions, solid waste, and any other waste.
- Locations of liquid effluent discharge points.
- Information on the quality of the air emissions and liquid effluent after relevant treatment, together with a description of the treatment system or pollution control technology installed to achieve the specified waste characteristics.
- Environmental alterations during construction (for example, land grading, right-of-way clearance, road building).
- Project employment.
- Proposed organization of environmental management staff and associated training.

- Projected occupational conditions related to worker health (for example, noise levels and workplace air quality) and safety (for example, hazardous areas, operations, or equipment).
- Proposed occupational health and safety programs, including training.

Description of the Baseline Environment

Detailed descriptions of baseline environmental conditions in and surrounding the operation must be provided. Graphic presentations of data (for example, figures and tables) are preferred where practical. Environmental features to be described (focusing on those environmental features likely to be affected by the project) include

- Climate and air quality.
- Landform (topography, geology, seismicity, and soils).
- Hydrology, water quality, and groundwater resources.
- Ecology, flora, and fauna (including identification of unique or sensitive natural habitats and locally or internationally recognized endangered or threatened species).
- Existing land and water resource uses, including potentially affected areas adjacent to the project site and focusing particularly on areas and uses likely to be affected (for example, receiving waters for liquid effluents).
- Archeological, historical, and culturally sensitive resources at the site.
- Socioeconomic conditions, especially potentially affected or concerned communities and organizations.

If the project is an expansion, modernization, or restart of existing activities, the EA must provide a summary description of any significant, persistent environmental problems related to past or current operations of the industrial facilities. This portion of the EA should identify any areas in which current conditions do not comply with applicable national, local, or World Bank requirements and guidelines and should identify measures required to bring these areas into compliance.

Potential Impacts of the Proposed Project

The EA must

- Provide a brief comparison of the impacts associated with alternative sites or processes, as appropriate, and key factors in decisions to select the proposed site and process.
- Identify all significant environmental, socioeconomic, and human

health and safety impacts associated with the project.

- Distinguish between beneficial and adverse impacts, direct and indirect impacts, and immediate and long-term impacts on the environment.
- Clearly identify significant impacts that are unavoidable or irreversible.
- Wherever possible, describe impacts quantitatively, in terms of environmental costs and benefits, assigning economic values where feasible.
- Characterize and explain significant information deficiencies and ensuing uncertainties associated with predictions of impact.
- Describe the planned and implemented programs, or both, to inform concerned governmental and nongovernmental organizations as well as local communities and involve them in the selection and design of mitigation, compensation, and monitoring measures.

In certain cases, cumulative impacts may need to be assessed. Such cases could arise when the project includes plans for significant future expansion or has the potential to interact with other proposed development activities such that the resulting impact on the environment is much greater than it would be if each occurred alone. As a general guidance, preparers of the EA should determine to the extent practical whether other development projects are being considered for the project area that are likely to be implemented and that could interact substantively with the proposed project.

Environmental Management and Mitigation Plan

The EA must include proposals for mitigation of any significant adverse impacts of the project and plans for ongoing management of the project to ensure that environmental impacts will be kept to a minimum throughout the lifetime of the project. This portion of the EA should

- Identify feasible and cost-effective measures to prevent significant adverse impacts or reduce them to acceptable levels.
- Estimate capital and recurrent costs for environmental management activities.
- Identify institutional and training requirements (that is, development of capabilities needed to implement mitigation measures and monitoring programs).
- Propose work plans and schedules (to ensure that actions are taken in phase with appropriate project engineering activities).
- Specify compensatory measures where mitigative measures are not feasible or cost-effective.

Monitoring Plan

The EA must include a detailed plan for monitoring waste discharges, emissions, and environmental parameters in and around the project. The proposed monitoring scheme must be adequate to allow determination of rates and concentrations of emissions and waste discharges from the project, occupational health and safety conditions, and the effectiveness of mitigating measures.

The monitoring plan should include an estimate of capital and operating costs and a description of other inputs (such as training and institutional strengthening) required for its implementation. The plan should also identify action levels for each parameter monitored, at which point action will be taken to reduce or mitigate associated impacts.

Specific elements to be included in this plan will vary according to the potential environmental and occupational health and safety concerns associated with each project. In many projects, elements of a monitoring program include

- Stack emissions and ambient air quality.
- Effluents released to surface waters, including storm water runoff and receiving water quality.
- Accident frequency and severity, temperature, noise, and air quality conditions at work stations.
- Socioeconomic conditions.

Coordination with Government Agencies, Affected Communities, and Nongovernmental Organizations

As mentioned earlier, the sponsors should make an active effort during the preparation of the EA to consult in a culturally acceptable context with appropriate government agencies and obtain the views of local NGOs, affected communities, and other affected groups. Records should be kept of meetings and other activities, communications, comments, and actions taken or project modifications made in response to public and community input. The EA should include a summary of steps taken to consult with local interested parties, highlights of key concerns of local interested parties, and a brief explanation of how these were addressed in the draft EA and project design.

References

All information sources referenced in the text should be identified, with full citations in the case of published information. In the case of unpublished documents and personal communications, sufficient information should be provided to inform the reader about the sources of information. For example, references to personal communications should indicate the names and affiliations of communicating parties and the approximate date of the communication. References to unpublished reports should include the name and location (for example, city and country) of the offices where such documents are located.

Glossary

936 funds: Bonds issued in conformity with a section in the U.S. tax code that permits a tax exemption on earnings of U.S. corporations in Puerto Rico, provided that the earnings are reinvested in Puerto Rico, Jamaica, or certain other Caribbean countries. Because of the reinvestment requirement, these funds carry interest rates below *LIBOR*. The U.S. Congress has recently voted to end 936 funds.

Advance payment guarantee: Assists the contractor in purchasing and assembling the materials, equipment, and personnel necessary to begin construction, so as to meet the requirements for receipt of progress payments under the contract.

American Depository Receipt (ADR): A certificate of ownership, issued by a U.S. bank, representing a claim on underlying foreign securities. ADRs may be traded in lieu of trading in the actual underlying shares.

Associated gas: Natural gas found dissolved in or together with crude oil in a reservoir.

Average levelized cost (average discounted cost): Measures the average capital and fuel costs for each type of a generating plant by dividing the present value of the cost stream by the present value of the output stream (under some circumstances called average incremental cost).

Balance of payments: The measurement of all international financial transactions between residents of a given country and foreign residents.

Balance of trade: The measurement of the foreign exchange value of exports minus imports for a given country.

Barrel (bl; petroleum): A unit of volume equal to 42 U.S. gallons.

Basis swap: An agreement in which one counterparty agrees to pay a floating price based on a commodity index at a certain delivery point in exchange for the floating price on the commodity futures market.

Bid bond: A bond required of bidders at the outset of a project to ensure that each bidder is serious and would accept the award of the contract if offered.

Bill of lading (B/L): A contract between a common carrier and a shipper to transport goods to a named destination. The bill of lading is also a receipt for the goods.

Black market: An illegal foreign exchange market.

BOO/BOOT/BLT schemes: Build-own-operate (BOO) and build-own-operate-transfer (BOOT) schemes are methods by which private sector participation in the energy sector is encouraged. Under these approaches, a project company under private ownership, or a joint venture with a minority public participation, is set up to plan, finance under limited recourse, design, construct, and operate energy generation facilities. In a BOOT arrangement, ownership of the facility is ultimately transferred to another entity after a specified period of operation. A variant is the Build-Lease-Transfer (BLT) scheme.

Branch: A foreign operation not incorporated in the host country, in contradistinction to a subsidiary.

British thermal unit (Btu): The quantity of heat needed to raise the temperature of 1 pound of water by 1°F at or near 39.2°F. (See *Heat content of a quantity of fuel, gross, and Heat content of a quantity of fuel, net.*)

Buy-back rate: The rate charged by a utility to a private power developer for buying back some or all the power the developer is obligated to deliver.

Capacity factor: The ratio of the electrical energy produced by a generating unit for a given period of time to the electrical energy that could have been produced at continuous full-power operation during the same period.

Capital account: That portion of the balance of payments that measures public and private international lending and investment.

Capital budgeting: The analytical approach used to determine whether investment in long-lived assets or projects is viable.

Capital markets: The financial markets in various countries in which different types of long-term debt and/or ownership of securities, or claims on those securities, are purchased and sold.

City gate: A point or measuring station at which a distribution gas utility receives gas from a natural gas pipeline company or transmission system.

Commercial risk: Comprises risks under the control of the owner of the project and includes risks of project development, timely construction, efficient and economic operation and maintenance (see also *force majeure risk, political risk, project risk*).

Common financing: Financing raised for a project as a whole and backed by several undertakings of the sponsors and appropriate inter-partner default clauses. Common financing is often used in the power sector (and is distinct from individual financing, usually used in oil projects, in which each partner is individually responsible for financing).

Concession: An arrangement whereby a private party leases assets for service provision from a public authority for an extended period and has responsibility for financing specified new fixed investments during the period; the assets revert to the public sector at expiration of the contract.

Concessional loan (or concessional financing): (See *soft loan*).

Consolidation: In the context of accounting for multinational corporations, the process of preparing a single "reporting currency" financial statement that combines financial statements of affiliates that are in fact measured in different currencies.

Cost and freight (C&F): Price, quoted by an exporter, that includes the cost of transportation to the named port of destination.

Cost risk: Includes the risks of cost increases caused by inflation, fluctuation of interest and foreign-exchange rates, changes in availability of materials and fuels, cost overruns, and project delays (combines with *revenue risk* into *margin risk*).

Cost, insurance, and freight (CIF): Exporter's quoted price including the cost of packaging, freight or carriage, insurance premium, and other charges paid in respect of the goods from the time of loading in the country of export to their arrival at the named port of destination or place of transshipment.

Counterguarantee: A third-party guarantee that is used when one party purchases a service or a product for monetary payments and this party's ability to make payments is in question. This guarantee is required by the provider of the service or product to ensure that payments will be made.

Creeping expropriation: (See *expropriation*).

Cross-currency swap: A transaction in which two counterparties agree to exchange principal and interest denominated in different currencies based on an agreed-upon currency exchange rate.

Cross-subsidies: The allocation of funds provided by one or more products or sectors of the economy to other products or sectors of the economy. Often this process is not transparent (for example, high prices for industrial users of electricity can be used to provide subsidies to domestic consumers).

Cubic foot (natural gas): A unit of volume equal to 1 cubic foot at a pressure base of 14.73 pounds standard per square inch absolute and a temperature base of 60° F.

Current account: In the balance of payments, the net flow of goods, services, and unilateral transfers (such as gifts) between a country and all foreign countries.

Debt service: Periodic payment of principal and interest on loans, bonds, or fixed- or floating-rate notes.

Depletion premium: Used in calculating the economic cost of supply of an exhaustible resource and indicates the opportunity cost of consuming a unit of the resource now rather than at a given point in the future, taking into account the cost of alternative fuels. In a country rich in the given resource, the depletion premium is negligible; in a country already short of the resource, the point of switching to an alternative fuel becomes the present, and the economic analysis is based on the *netback value*.

Devaluation: A government action to reduce the purchasing power or value of local currency against convertible currencies.

Development well: A well drilled within the proved area of an oil or gas reservoir to the depth of a stratigraphic horizon known to be productive.

Economic analysis: An analysis that takes into account the costs and benefits of a project to society or the country as a whole. For example, if electricity is subsidized, an economic analysis would use as its basis the price paid by the consumer plus the subsidy (compare *financial analysis*).

Electric power plant: A station containing prime movers, electric generators, and auxiliary equipment for converting mechanical, chemical, and/or fission energy into electric energy.

Electrical system energy losses: The amount of energy lost during generation, transmission, and distribution of electricity, including plant and unaccounted-for uses.

Electricity generation, gross: The total amount of electric energy produced by the generating station or stations, measured at the generator terminals.

Electricity generation, net: Gross generation less electricity consumed at the generating plant for station use. Electricity required for pumping at pumped-storage plants is regarded as plant use and is deducted from gross generation.

Electricity generation: The process of producing electric energy or transforming other forms of energy into electric energy. Also the amount of electric energy produced or expressed in watt-hours (Wh).

End-use sectors: The residential, commercial, industrial, and transportation sectors of the economy.

Escrow account: A special account, often outside the country hosting the project, to which all or part of project revenues are channeled to ensure disbursement of funds according to agreed conditions and priorities. When all funds are deposited, the disbursement first covers the operating cost and then debt service before remitting the remainder to the project company. An escrow account agreement terminates when all debt obligations have been fully paid.

Eurobond: A bond sold outside the country in whose currency it is denominated.

Eurocurrency: A currency deposited in a bank located in a country other than the country issuing the currency.

Eurodollar: A U.S. dollar deposited in a bank outside the United States.

European Currency Unit (ECU): Composite currency created by the European Monetary System to function as a reserve currency numeraire for denominating a number of financial instruments.

Exchange rate: The price of a unit of one country's currency expressed in terms of the currency of some other country.

Export parity: Price minus associated transportation cost.

Expropriation: A forced transfer of ownership from a private owner to a government institution ("creeping" expropriation may result from the growth of a *host country's* take through increased taxes, royalties or fees during project construction or operation).

Financial analysis: An analysis that takes into account the costs and benefits of a project to the entity undertaking it For example, if electricity is subsidized, the financial analysis would consider only the revenue to the entity and not the overall benefit of any subsidy given to the buyer.

Financial closing: Occurs when all the conditions of lenders and investors have been met, and financing disbursements can take place.

Financial closure: The point at which the legal and financial commitments for a project are formally executed; generally followed soon after by disbursement of funds and beginning of project construction.

Financial derivatives: Instruments such as interest rate swaps and caps as well as currency swaps and options used to hedge against unprotected increases in financial costs, particularly those caused by an increase in interest rates or an adverse movement in the exchange rate.

Financial engineering: The complex discipline of combining various instruments for securing guarantees, borrowing, and mobilizing equity toward the end of financing a project.

Fixed exchange rate: Foreign exchange rate set and maintained by government support.

Floating exchange rate: Foreign exchange rates determined by demand and supply in an open market that is presumably free of government interference.

Force majeure risk: Danger from unavoidable events such as tornadoes, earthquakes, hurricanes, and lightning that (combined with *political risk* and *commercial risk*) constitutes part of total *project risk*.

Forward foreign exchange contract: A contract to pay or receive specific amounts of a currency at a future date in exchange for another currency at an agreed-upon exchange rate.

Forward rate agreement: An agreement to exchange dollar amounts at a specified future date based on the difference between a particular interest rate index and an agreed-upon fixed interest rate.

Fossil fuel: Any naturally occurring organic fuel, such as petroleum, coal, and natural gas.

Franchise: The grant of certain rights to an individual group, partnership, or corporation (sometimes called a *concession*).

Free on board (FOB): International trade term in which the exporter's quoted price includes the cost of loading goods into transport vessels at a named point.

Futures contract: A standardized contract traded on an organized exchange that obligates one party to buy and another party to sell a specific asset at a future date.

Gearing (also called *debt ratio* or *leverage*): The ratio of a project's fixed-interest debt to its equity plus debt. A highly geared (or leveraged) project is one in which the proportion of debt is comparatively great. Energy projects are typically financed in a range of 20 to 40 percent equity and 60 to 80 percent debt. A project with a 30 percent equity component thus would have a gearing of 70 percent. Lenders tend to prefer lower gearing, which means a greater contributions of equity and hence greater commitment by project sponsors and stockholders, whereas project sponsors are likely to prefer higher gearing to minimize the capital they lock into the project.

Gross domestic product (GDP): Measures the total value of goods and services produced within a country over a given period, usually a year.

Gross domestic product (GDP): The total value of goods and services produced by labor and property located in a country.

Gross national product (GNP): Comprises GDP plus income to domestic entities from foreign holdings minus income to foreign entities from domestic holdings.

Gross product worth (GPW): The weighted average value of the refinery yield; that is the mix of refined products that are derivable from a given quantify of crude. GPW is obtained by multiplying the prevailing price of the derived products by their percentage yield from the crude product (cf. *Net product worth, netback value*).

Hard currency: All major convertible currencies, such as the U.S. dollar, the British pound, the German mark, the Japanese yen, the French franc, the Swiss franc, the Italian lira, and the Dutch guilder.

Heat content of a quantity of fuel, gross: The total amount of heat released when a fuel is burned. Coal, crude oil, and natural gas all include chemical compounds of carbon and hydrogen. When those fuels are burned, the carbon and hydrogen combine with oxygen in the air to produce carbon dioxide and water. Some of the energy released in burning goes into transforming the water into steam and is usually lost. The amount of heat spent in transforming the water into steam is counted as part of gross heat content but is not counted as part of net heat content.

Heat content of a quantity of fuel, net: The amount of usable heat energy released when a fuel is burned under conditions similar to those in which it is normally used. Also referred to as the lower heating value.

Heavy oil: The fuel oils remaining after the lighter oils have been distilled off during the refining process. Except for startup and flame stabilization, virtually all petroleum used in steam-electric power plants is heavy oil.

Home country: The country in which a private energy developer is registered.

Host country: The country in which the energy project is taking place.

Hydrocarbon: An organic chemical compound of hydrogen and carbon in the gaseous, liquid, or solid phase. The molecular structure of hydrocarbon compounds varies from the simplest (methane, the primary constituent of natural gas) to the very heavy and very complex.

Hydroelectric power: The production of electricity from the kinetic energy of falling water.

Implementation agreements: Project-specific agreements that provide government assurances and guarantees to private energy producers required for successful project development and allocation of risk.

Implementation team: The private energy developer and its contractors.

Independent power producer (IPP): Private energy producer who has developed power plants, typically on a project finance basis, to sell power to an existing utility or directly to distributors or large consumers.

Indexed tariff: An adjusted tariff, based on a variable such as periodic fuel price, interest rates (local or foreign), exchange rate or inflation rate.

Interest rate swap: An agreement in which one counterparty agrees to pay a fixed rate of interest to the other counterparty in exchange for a variable rate of interest on a fixed notional principal amount over a specified period of time.

Internal combustion electric power plant: A power plant in which the prime mover is an internal combustion engine. Diesel or gas-fired engines are the principal types used in electric power plants. The plant is usually operated during periods of high demand for electricity.

Internal rate of return (IRR): The discount rate that equates the present value of a project's expected cash inflows to the present value of the project's expected costs.

International oil company (IOC): A category used to refer to integrated major oil companies, large independents, and large exploration and production companies. The category does not include national oil companies (NOCs).

Investors: Individuals, groups, or companies that invest cash in a private energy developer, group, or company.

Irrevocable letter of credit: A guarantee by a commercial bank to provide credit on demand up to a specified maximum limit. This service is provided by banks for a fee.

Joint venture: A business venture that is owned by two or more other business ventures. Often the several business owners are from different countries.

Kabushiki-Kaishi (KK): Japanese term for stock company.

Letter of credit (L/C): An instrument issued by a bank, in which the bank promises to pay a beneficiary upon presentation of documents specified in the letter of credit.

Limited-recourse financing: A lending arrangement under which repayment of a loan and recourse in the event of a default relies mainly on the project's cash flow, but project sponsors remain responsible to cover default during a limited period of time.

Liquefied natural gas (LNG): Natural gas (primarily methane) that has been liquefied by reducing its temperature to 260° F at atmospheric pressure.

Liquefied petroleum gases (LPG): Ethane, ethylene, propane, propylene, normal butane, butylene, and isobutane produced at refineries or natural gas processing plants, including plants that fractionate new natural gas plant liquids.

London Interbank Offer Rate (LIBOR): The deposit rate applicable to interbank loans within London.

Lump-sum turnkey (LSTK) contract: Used to ensure that the project is completed on time, within budget, and with acceptable standards of operation. These contracts involve a set sum for completion of a project but have stringent reward and penalty clauses that protect sponsors against cost overruns and operational deficiencies (see also *turnkey contract*).

Maintenance bond: Provides a source of funds for correcting construction or performance defects discovered after completion of project construction. Typically, the *performance bond* and the *retention bond* are converted to maintenance bonds on completion of the contract.

Margin risk: The sum of *cost risks* and *revenue risks.*

Mixed credits: Blending of credits from different sources and at different levels of *concessional* finance.

Naphtha: A generic term applied to a petroleum fraction with an approximate boiling range between 122 and 400° F.

National oil company (NOC): A category used to refer to state-owned oil companies.

Natural gas plant liquids (NGPL): Natural gas liquids recovered from natural gas in processing plants and, in some situations, from natural gas field facilities, as well as those extracted by fractionators.

Natural gas, dry: The marketable portion of natural gas production, which is obtained by subtracting extraction losses, including natural gas liquids removed at natural gas processing plants, from total production.

Natural gas, wet: Natural gas prior to the extraction of liquids and other miscellaneous products.

Natural gas: A mixture of hydrocarbons (principally methane) and small quantities of various nonhydrocarbons existing in the gaseous phase or in solution with crude oil in underground reservoirs.

Net present value (NPV): A capital budgeting approach in which the present value of expected future case inflows in subtracted from the present value of outflows to determine the "net" present value.

Net product worth (NPW): Derived by subtracting from the *gross product worth* of a given quantity of oil or gas the operating expenses involved in handling the last (marginal) quantity of the crude product (amortization and depreciation are not included).

Netback value: An estimate of the value of products that a given quantity of a specific type of crude oil or natural gas will yield after passing through a particular configuration of refinery or processing plant. It is obtained by subtracting the *net product worth* from the *gross product worth*. Further subtracting the costs of transportation and insurance gives the netback value at the port of loading.

Nonrecourse financing: Recourse for debt repayment, default, or both belongs exclusively to the project company.

Nuclear electric power: Electricity generated by an electric power plant whose turbines are driven by steam generated in a reactor by heat from the fissioning of nuclear fuel.

OECD Consensus: Guidelines intended to prevent distortions in price competition among manufacturers of different countries. The OECD Consensus is derived from an OECD agreement, "Arrangement on Guidelines for Officially Supported Credits" (1978), which limits export credit to 85 percent of the contract value and holds interest rates to a minimum of the OECD interest rate matrix, which is revised. semi-annually.

Oil well: A well completed for the production of crude oil from one or more oil zones or reservoirs. Wells producing both crude oil and natural gas are classified as oil wells.

Onlending: Lending through financial intermediaries (also called relending).

Option: A contract that provides the option holder the right, but not the obligation, to buy or sell the underlying instrument.

Organisation for Economic Co-operation and Development (OECD): Current members are Australia, Austria, Belgium, Canada, Denmark, Finland, France, Germany, Greece, Iceland, Ireland, Italy, Japan, Luxembourg, the Netherlands, New Zealand, Norway, Portugal, Spain, Sweden, Switzerland, Turkey, the United Kingdom, the United States and its territories (Guam, Puerto Rico, and the Virgin Islands).

Organization of Petroleum Exporting Countries (OPEC): Countries that have organized for the purpose of negotiating with oil companies on matters of oil production, prices, and future concession rights. Current members are Algeria, Gabon, Indonesia, Iran, Iraq, Kuwait, Libya, Nigeria, Qatar, Saudi Arabia, the United Arab Emirates, and Venezuela.

Parallel financing: Arranging separate loan agreements for each type of financing in support of a project.

Performance bond: Provides additional funds in the event a contractor fails to perform for any reason. The existence of such a bond is also an endorsement of the credit and confidence of the guarantor in the ability and professional standing of the contractor.

Performance bonds: Guarantees purchased by the project developer issued by commercial banks or insurance companies for an entity to guarantee full and successful implementation of a contract according to prespecified performance guidelines.

Petrochemical feedstocks: Chemical feedstocks derived from petroleum principally for the manufacture of chemicals, synthetic rubber, and a variety of plastics.

Petroleum products: Products obtained from the processing of crude oil (including lease condensate), natural gas, and other hydrocarbon compounds. Petroleum products include unfinished oil, liquefied petroleum gases, pentanes plus, aviation gasoline, motor gasoline, naphtha-type jet fuel, kerosene-type jet fuel, kerosene, distillate fuel oil, residual fuel oil, petrochemical feedstocks, special naphthas, lubricants, waxes, petroleum coke, asphalt, road oil, still gas, and miscellaneous products.

Petroleum: A generic term applied to oil and oil products in all forms, such as crude oil, lease condensate, unfinished oils, petroleum products, natural gas plant liquids, and nonhydrocarbon compounds blended into finished petroleum products.

Pipeline fuel: Gas consumed in the operation of pipelines, primarily in compressors.

Plant downtime: Time when a plant is not producing power because of scheduled or forced outage or shutdown.

Political risk: Presumably avoidable risks (in contrast to *force majeure risk*) to a project stemming from political uncertainty or difficulty in the host country and including expropriation of project property, breach of undertakings by the host government or government entities, civil unrest or revolution, war, prevention of profit repatriation, and inconvertibility of local currency.

Power purchaser: The entity purchasing power from a private power developer. Usually, the public utility of the host country is the power purchaser.

Price swap: An agreement in which one counterparty agrees to pay a fixed price for a commodity to the other counterparty in exchange for a price based on a commodity index. The payments are based on fixed notional quantities.

Private power developer: An individual group or company that develops power plants on a private basis to own, operate, lease or transfer.

Project company: The special-purpose entity that assumes legal and financial responsibility for construction and operation of the project. Recourse is limited to the project company.

Project risk: The total risk, including commercial, political and *force majeure risks.*

Put option: Allows holders to present or "put" bonds for repayment at maturity.

Put-or-pay contract (also called *supply-or-pay contract*): Provided by suppliers of energy, raw materials, or products to projects needing assured inputs over long periods, at predictable prices, to meet production cost targets. The put-or-pay obligor must either supply the energy, raw material, or product or pay the project company the difference in cost incurred in obtaining the input from another source.

Quasi-equity: A special type of funding meant to attract risk-averse investors to a project, it is issued in the form of preferred shares, whose holders are paid before holders of ordinary shares but after creditors and holders of debt financing.

Refinery (petroleum): An installation that manufactures finished petroleum products from crude oil, unfinished oils, natural gas liquids, other hydrocarbons, and alcohol.

Relending: (See *onlending*).

Retention money bond: A bond that contractors substitute for the portion of the progress payment to a contractor that sponsors would otherwise retain to cover unforeseen expenses caused by any contractor mistakes in construction.

Revenue risk: Risk to a project from fluctuations in demand and price (combines with *cost risk* into *margin risk*)

Risk profile: The level of risk attributable to political, economic, or financial uncertainty to which an investor is exposed. This determines the rate of return that an investor requires in order to tolerate exposure to the level of adversity in any country or project.

Rule 144A: Allows the sale of restricted investment-grade securities (notes or bonds) to "qualified institutional buyers" without registration with the U.S. Securities and Exchange Commission, thereby providing foreign companies access to the U.S. capital market. The issue of these securities does not require the same detailed financial information as a public offering. However, three years after their issue, 144A securities can be freely traded in the U.S. market.

Senior loan: A loan having priority in repayment over other loans or obligations.

Soft loan (or soft financing): A loan or financing arrangement bearing no interest or interest below the cost of the capital lent. Loans of the World Bank are somewhat below commercial rates but are *not* "soft." The World Bank Group affiliate, the International Development Association, lends to the world's poorest countries on "soft" or concessional terms.

Sovereign guarantee: Government guarantee(for example, guarantee of repayment of a loan by a state entity or of the obligations of a purchasing utility under a power purchase agreement.

Sovereign rating: Measures the ability and willingness of a country to service its debts; credit rating agencies may take a country's sovereign rating as representing a ceiling for the rating of any other entity domiciled in the country, recognizing that the central government's control over economic, fiscal, monetary, and trade policy renders the country's credit standing above that of any other debtor in the country.

Sponsors' completion guarantee: A guarantee that commits the sponsors to completing the project within a certain time period and to providing funds to pay all cost overruns. However, other guarantees (such as *performance bonds* and *maintenance bonds* by contractors, equipment suppliers, and so on) are normally sought to protect the project sponsors and to provide financiers with additional assurances that project completion risks are managed.

Synthetic natural gas (SNG): A manufactured product chemically similar in most respects to natural gas, resulting from the conversion or reforming of petroleum hydrocarbons. It may easily be substituted for, or interchanged with, pipeline quality natural gas. Also referred to as substitute natural gas.

Take-or-pay contract: An unconditional contractual arrangement between the project sponsor and project's customer obliging the customer to make periodic payments in the future in return for fixed amounts or quantities of products, goods, or services at specified prices, whether or not the service or the product is used by the customer. Payments are usually subject to escalation because of inflation. The obligor can protect its interest by retaining rights to take over the project in the event of failure by the supplier to perform, but in doing so, the obligor would also have to assume or pay the debt used to finance the project.

Tax holidays: Exemptions from some or all taxes for a specified period.

Throughput contract (sometimes called a *tolling agreement*, a *cost-of-service tariff*, or a *deficiency agreement*): The equivalent to *take-or-pay contracts* for projects that provide services such as power transmission, power distribution, oil pipeline transportation, or refining in which he obligor pays whether the service is used or not. Because the guarantee is unconditional and lasts for the life of the loan, lenders regard throughput contracts as a guaranteed source of income.

Turnkey contract: A contract given by the project developer to a prime contractor who will be responsible for design and implementation of a project from start to finish, and who will provide a completed, operational project on a stipulated date, on a lump-sum turnkey (LSTK) basis or at cost plus a fixed fee.

Vented natural gas: Gas released into the air on the base site or at processing plants.

Wellhead price: The value of crude oil or natural gas at the mouth of the well.

Bibliography

Bankes-Jones, Tony, and Margaret Gossoling. 1994. "Energy and Project Finance." *Petroleum Economist* (June).

Baughman, David, and Matthew Buresch. 1994. "Mobilizing Private Capital for the Power Sector: Experience in Asia and Latin America." Joint World Bank–USAID Discussion Paper. World Bank Cofinancing and Financial Advisory Services Department, World Bank Industry and Energy Department, and USAID Office of Energy, Environment and Technology, Washington, D.C., November.

Berry, Colin. 1991. "Criteria for Successful Project Financing." In *Project Finance Yearbook (1991/92)*, pp. 15–20. London: Euromoney Publications.

Besant-Jones, John, ed. 1990. "Private Sector Participation in Power through BOOT Schemes." IEN Energy Series Paper 33. World Bank, Industry and Energy Department, Washington, D.C.

Bond, Gary, and Laurence Carter. 1995. "Financing Energy Projects: Experience of the International Finance Corporation." *Energy Policy* 23 (December). Forthcoming.

Bond, Gary, and Laurence Carter. 1994. *Financing Private Infrastructure Projects: Emerging Trends from IFC's Experience.* IFC Discussion Paper 23. Washington, D.C.: International Finance Corporation.

British Petroleum Company. 1995. *BP Statistical Review of World Energy.* London.

British Petroleum Company. 1995. *BP Review of World Gas*. London.

Brooks, Andrew. 1994. "The Credit Agency's Perspective." Paper presented at the Annual Oil Project Finance Conference in London (June). Processed.

Cambridge Energy Resources Association, Inc. (CERA) & Arthur Andersen & Co. 1994. *World Oil Trends*. Cambridge, Mass.

CEDIGAZ. Annual. *Natural Gas in the World: (Year) Survey*. Paris.

Clifford, Chance. 1991. *Project Finance*. London: IFR Books.

Crawford, J. P. S. "Infrastructure Project Finance in Developing Countries." In *Project Finance Yearbook (1991/92)*, pp. 84–94. London: Euromoney Publications.

Dryland, Hugo. 1991. "Natural Resource Finance in the LDC Context." In *Project Finance Yearbook (1991/92)*, pp. 107–11. London: Euromoney Publications.

Dunkerley, Joy. 1995. "Financing the Energy Sector in Developing Countries: An Overview." *Energy Policy* 23 (December). Forthcoming.

Eiteman, David, and Arthur Stonehill. 1989. *Multinational Business Finance*. New York: Addison-Wesley.

Energy Information Administration, U.S. Department of Energy. 1995. *Country Analysis Briefs*. Washington, D.C.

Energy Information Administration, U.S. Department of Energy. 1995. *International Energy Outlook*. Washington, D.C.

Fabizzi, Frank, and Dessa Fabizzi. 1995. *The Handbook of Fixed Income Securities*. New York: Irwin Publishing.

Fairclough, A. J. 1994. *World Development AID and Joint Venture Finance 1994–1995*. London: Kensington Publications.

Goldstein, Morris, and Landau Folkers. 1993. *International Capital Markets; Part I and Part II*. Washington, D.C.: International Monetary Fund.

Håmsø, Bjørn, Afsaneh Mashayekhi, and Hossein Razavi. 1994. "The International Gas Trade: Potential Major Projects." *Annual Review of Energy and Environment* 19:37–73.

Harvey, Charles. 1983. *Analysis of Project Finance in Developing Countries*. London: Heinemann.

Heus, Reinhold. 1994. "Outlook For the Energy Sector of Russia and Central Asia: Structural and Cultural Perspectives." In Thomas W. Wälde, George K. Ndi, eds., *International Oil & Gas Investment: Moving Eastward?*, pp. 227–34. London: Graham & Trotman.

Himberg, Harvey. 1995. "OPIC Support to Power Projects in Asia." *Energy Policy* 23 (December). Forthcoming.

Hollis, Sheila Slocum, and John W. Berresford. 1994. "Structuring Legal Relationships in Oil and Gas Exploration and Development in Frontier Countries." In Thomas W. Wälde, George K. Ndi, eds., *International Oil & Gas Investment: Moving Eastward?*, pp. 29–59. London: Graham & Trotman.

Humphries, Michael. 1993. *The Competition for Capital in the International Oil and Gas Industry*. Washington, D.C.: The Petroleum Finance Company.

Humphries, Michael. 1995. "The Competitive Environment for Oil and Gas Financing." *Energy Policy* 23 (December). Forthcoming.

Hurt, Christopher. 1988. "Financing Natural Gas Projects in Developing Countries." In Thomas Wälde and Nicky Beredjick, eds., *Petroleum Investment Policies in Developing Countries*, pp. 189–203. London: Graham & Trotman.

Ingham, Richard. 1991. "Risk Management." In *Project Finance Yearbook (1991/92)*, pp. 35–42. London: Euromoney Publications.

International Energy Agency. Annual. *Energy Statistics and Balances of Non–OECD Countries (Year).* Paris.

International Energy Agency. Annual. *World Energy Outlook (Year).* Paris.

International Energy Agency. 1994. *Energy in Developing Countries.* Paris.

International Energy Agency, Diversification Division. 1994. *Electricity Supply Industry: Structure, Ownership and Regulation in OECD Countries.* Paris.

International Monetary Fund. Annual. *International Capital Markets (Year).* Washington, D.C.

International Monetary Fund. Annual. *International Financial Statistics (Year).* Washington, D.C.

International Monetary Fund. 1995. *Private Market Financing for Developing Countries.* World Economic and Financial Surveys. Washington, D.C.

Jechoutek, Karl G., and Ranjit Lamech. 1995. "New Directions in Electric Power Financing." *Energy Policy* 23 (December). Forthcoming.

Julius, DeAnne S., and Afsaneh Mashayekhi. 1990. *Economics of Natural Gas: Pricing, Planning and Policy.* Oxford: Oxford University Press.

Kayaloff, Isabelle. 1988. *Export and Project Finance.* London: Euromoney Publications.

Konoplyanik, A. 1994. "The Russian Oil Industry and Foreign Investments: Legal Aspects and the Problem of Business Risk." In Thomas W. Wälde, George K. Ndi, eds., *International Oil & Gas Investment: Moving Eastward?*, pp. 181–194. London: Graham & Trotman.

McKechnie, Gordon. 1983. *Energy Finance.* London: Euromoney Publications.

Miranda, Carlos. 1994. "Financing of Gas Pipelines in a Less Developed Country Two Decades Apart." Paper presented at the 19th World Gas Conference, Milan, Italy, June 20–23.

Nevitt, Peter K. 1989. *Project Financing*. London: Euromoney Publications.

Onorato, William T. 1995. "Legislative Frameworks Used to Foster Petroleum Development." World Bank Policy Research Working Paper 1420. Washington, D.C.

Parker, Nicholas. 1993. *Investing in Emerging Economies: A Business Guide to Official Assistance and Finance*. London: Economist Intelligence Unit.

PennWell Publishing. Annual. *International Petroleum Encyclopedia*. Tulsa, Okla.: PennWell.

Perkins, Patricio. 1994. "Financing of Gas Projects in Developing Countries under a Privatization Scheme: The Case in Argentina." Presented at the 19th World Gas Conference in Milan, Italy, June 20–23.

Petroleum Economist. 1993. "New Directions in Energy Finance." Petroleum Economist 60 (June): 4–5.

Petroleum Intelligence Weekly. 1994. "PIW Ranks the World's Top 50 Oil Companies." *Petroleum Intelligence Weekly*, December 12, 1994. New York.

Pittore, Francesco. 1991. "The Export Credit Agency Approach to Project Finance." In *Project Finance Yearbook (1991/92)*, pp. 49–52. London: Euromoney Publications.

Power Development, Efficiency and Household Fuels Division, The World Bank. 1995. "Power Sector Performance Monitoring Indicators." Draft paper. World Bank Industry and Energy Department, Washington, D.C. May.

Razavi, Hossein. 1995. "Oil and Gas Financing by the World Bank." *Energy Policy* 23 (December). Forthcoming.

Razavi, Hossein, and Fereidun Fesharaki. 1991. *Fundamentals of Petroleum Trading*. Westport, Conn.: Praeger.

Ryrie, Sir William. 1991. "Investing in Development." In *Project Finance Yearbook (1991/92)*, pp. 27–34. London: Euromoney Publications.

Smith, Arthur L. 1994. "Renaissance of the E&P Independents." *Petroleum Economist* (June): 80–84.

Svensson, Bent, and Patrick. A. Wright. 1994. "Financing of Natural Gas Projects in Developing Countries." Paper presented at the 19th World Gas Conference, Milan, Italy, June 20–23.

Syrett, Stephen. 1991. "The Future of Project Finance." In *Project Finance Yearbook (1991/92)*, pp. 11–14. London: Euromoney Publications.

Van Meurs, Pedro. 1988. "Financial and Fiscal Arrangements for Petroleum Development: An Economic Analysis." In Thomas Wälde and Nicky Beredjick, eds., *Petroleum Investment Policies in Developing Countries*, pp. 47–79. London: Graham & Trotman.

United Nations. Annual. *Energy Balances and Electricity Profiles (Year)*. New York.

United Nations. Annual. *Energy Statistics Yearbook (Year)*. New York.

World Bank. Annual. *Trends in Developing Economies* (Year). Washington, D.C.

World Bank. Annual. *World Debt Tables: (Year). External Finance for Developing Countries. Vol. I: Analysis and Summary Tables, and Vol. II: Country Tables*. Washington, D.C.

World Bank. Annual. *World Tables (Year)*. Baltimore, Md., and Washington, D.C.: Johns Hopkins University Press/World Bank.

World Bank. 1992. *Guide to International Business Opportunities*. Washington D.C.

World Bank. 1993. "Power Supply in Developing Countries." Proceedings from a roundtable co-sponsored by the World Bank and Electricité de France, April 27–28. Occasional Paper 1. Industry and Energy Department, Washington, D.C.

World Bank. 1995. *The Emerging Asian Bond Market.* East Asia and Pacific Region, Washington, D.C.

World Bank and USAID. 1994. "Submission and Evaluation of Proposals for Private Power Generation Projects in Developing Countries." Occasional Paper 2. Industry and Energy Department, Washington, D.C.

INDEX

A

B

C

D

E

F

G

H

I

J

K

L

M

N

O

P

Q

R

S

T

U

W

Y